CISM COURSES AND LECTURES

The series presents lecture notes, monographs, edited works and proceedings in the field of Mechanics, Engineering, Computer Science and Applied Mathematics.
Purpose of the series is to make known in the international scientific and technical community results obtained in some of the activities organized by CISM, the International Centre for Mechanical Sciences.

INTERNATIONAL CENTRE FOR MECHANICAL SCIENCES

COURSES AND LECTURES - No. 404

NEURAL NETWORKS IN THE ANALYSIS AND DESIGN OF STRUCTURES

EDITED BY

ZENON WASZCZYSZYN
CRACOW UNIVERSITY OF TECHNOLOGY

Springer-Verlag Wien GmbH

This volume contains 128 illustrations

© 1999 by Springer-Verlag Wien
Originally published by CISM, Udine in 1999.

SPIN 10745149

In order to make this volume available as economically and as
rapidly as possible the authors' typescripts have been
reproduced in their original forms. This method unfortunately
has its typographical limitations but it is hoped that they in no
way distract the reader.

ISBN 978-3-211-83322-3   ISBN 978-3-7091-2484-0 (eBook)
DOI 10.1007/978-3-7091-2484-0

Professor Panadis D. Panagiotopoulos

Member of CISM Scientific Council died untimely and suddenly on August 12, 1998. An outstanding scientist, enthusiastic organiser of scientific life, excellent teacher, warm and always friendly for his colleagues and associates in Greece and abroad was Prof. P.D. Panagiotpoulos a great personality of international range.

Panadis Panagiotopoulos was born in Thessaloniki on January 1, 1950. At the age of five years he was enrolled at the Experimental School of the Aristotle University, Thessaloniki. After his M.Sc. at the Department of Civil Engineering of the Aristotle University he joined his career with this University. After Ph.D Thesis defence in Thessaloniki in 1972 he spent five years at the RWTH Aachen as the Humboldt Research Fellow and then as Head of a research group at the Institute of Technical Mechanics. In 1977 he defended his habilitation at the Faculty of Mathematics and Physics of RWTH Aachen. Since this time he devoded his life and scientific activity between Thessaloniki and Aachen. In 1978 he was appointed Full Professor and Director of the Institute of Steel Structures at Department of Civil Engineering, Aristotle University, Thessaloniki. In 1981 he was nominated Honorary Professor of Mechanics at the Faculty of Mathematics and Physics of RWTH Aachen.

The main fields of Prof. P.D. Panagiotopoulos's interests were nonsmooth mechanics, hemivariational inequalities, inequality problems in mechanics, optimization methods in structural mechanics, applications of artificial neural networks and fractals in mechanics. As the only author and co-author he published 9 books in English (one of them was translated into Russian). He published more than 200 papers in international journals and in books of scientific institutions and more than 50 of his papers were printed in conference proceedings. He was the Visiting Professor in Universities of Hamburg, Namur, Rio de Janeiro, MIT Cambridge Mass., USA. He was the Full Member of the Academia Europea, Corresponding Member of the Academy of Athens. In 1995 he was Recipient of the International Price "Agostinelli" of the Academia Nazionale dei Lincei, Rome, Italy.

It is not possible to write down in this biosketch the full list of Prof. Panagiotopoulos activities. Let us only mention his participation in 10 Editorial Committees and membership of GAMM, ISIMM, ASCE, SIAM, Math. Programming Soc., DCEG, DRG, AMS, DGM, and in Greece of TCG, HSTAM, HSRMS, HSCM. And, moreover, he found time for engineering practice as a member of consulting groups of the Greek Ministry of Public Works and for the German automobile industry.

And at the end it should be emphasized his cooperation with CISM. He was not only a very active Member of Scientific Council but he participated in many courses and advanced schools as an outstanding lecturer.

It is unbelievable that one man, in such a short period of his activity could achieve so much!

Aris V. Avdelas
Zenon Waszczyszyn

PREFACE

The Artificial Neural Networks (ANNs) are introduced to many fields of science and technology as a new tool of data processing. In recent years ANNs have also been exten-sively applied to problems of mechanics. They enable us to deal with problems which are difficult or even impossible to be analysed by means of standard or parallel computers.

That is why it was quite obvious that neurocomputing and prospects of its applications in the analysis of mechanics problems drew Professor Sandor Káliszky's attention. After a discussion with him in September 1996 and then with Professor Panadis Panagiotopoulos I was encouraged to start with the arrangement of the CISM Advanced School on Neural Networks in Mechanics of Structures and Materials. It was my great pleasure that my invitation directed to potential lecturers was kindly accepted not only by Professor Panadis D. Panagiotopoulos of the Aristotle University of Thessaloniki, Greece but also by Professor William M. Jenkins of the University of Leeds, Barry H.V. Topping of Heriot-Watt University, United Kingdom, Prabhat Hajela, Rensselaer Polytechnic Institute, Troy NY, USA and Doctor Paolo Venini, University of Pavia, Italy.

The School was held during the week of October 19 to 23, 1998 in the beautiful CISM residence, Palazzo del Torso in Udine, Italy.

The book compiled after the School reflects its main ideas. They are related to discussions not only about the theoretical background of ANNs, but first of all about their applications in selected fields of structural mechanics.

The ANN fundamentals treat special types of neural networks, their architectures and algorithms. Relations to genetic algorithms and fuzzy systems are considered as well.

The applications of ANNs are devoted to various problems of mechanics, as well as to the analysis and design of structures. Applications of ANNs are discussed with respect to optimisation problems, identification and damage detection, plasticity and fracture mechanics, to active control of structures, to application of advanced methods of data processing.

The book is addressed to advanced graduate students, postgraduate and doctoral students, academic and research engineers and professionals working in the field of computational mechanics and structural engineering.

The book is devoted to Professor P.D. Panagiotopoulos who died untimely on August 12, 1998. I think that now this is the only possible gratitude to him since without his support and enthusiasm the CISM Advanced School could not have been arranged and this book could not have been written.

I wish to express my deepest gratitude to Professor Sandor Káliszky for his friendly support and to Professor Aris Avdelas for his help during my work with writing the Chapter related to Professor Panagiotopoulos. I wish to express my great thanks to all authors of this book for their effort with preparation of their Chapters. At last but not least I have to express my gratitude to Professor Carlo Tasso, the CISM Scientific Editor, for his instructions and patience during the preparation of the manuscripts.

Zenon Waszczyszyn

CONTENTS

Page

CONTENTS

CHAPTER 1

FUNDAMENTALS OF ARTIFICIAL NEURAL NETWORKS

Z. Waszczyszyn

Cracow University of Technology, Cracow, Poland

ABSTRACT

The introduction to this Chapter concerns principal ideas of the formulation of Artificial Neural Networks (ANNs), main features of neurocomputation, its development and applications. The main attention is paid to feedforward NNs, especially to the error backpropagation algorithm and Back-Propagation Neural Networks (BPNNs). Data selection and preprocessing, learning methods, BPNN generalisation, multilayered NN architectures and radial basis functions are discussed. In the frame of unsupervised learning different learning rules are listed and their usage in Kohonen self-organizing networks, ART (Adaptive Resonanse Theory) networks and CP (Counter-Propagation) networks are considered. The Hopfield discrete and continuous networks and the BAM (Bidirectional Adaptive Memory) network are discussed as examples of recurrent neural networks. The references cover theoretical background of ANNs, review papers and selected papers on applications of neurocomputing to the analysis of problems in mechanics and structural engineering.

1.1. INTRODUCTION

1.1.1 Biological and artificial neural networks

The nervous systems of living organisms are in general composed of three parts (subsystems), cf. Fig.1.1:
- central nervous system,
- peripheral nervous system,
- autonomic nervous system.

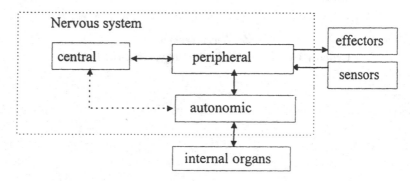

Fig.1.1: Scheme of nervous system

The central nervous system contains the brain and marrow. It is an enormous network composed of neural units, connections and joints. The main function of this system is to control the activity of the whole organism basing on information processing. The information signals are transmitted along the peripheral nervous system which has contact with the external sensors and effectors. The autonomic nervous system supervises the activity of the internal organs.

The most sophisticated part of the nervous system is the brain. It can be considered as a highly complex, nonlinear and parallel information-processing system. The basic elements of the brain are the neural cells called *neurons*. In Fig.1.2a the main parts of a neuron are shown.

The neuron receives signals and produces a response. The neuron transmits information using electrical signals. The processing of this information involves a combination of electrical and chemical processes. When an electrical impulse arrives at a synapse, it involves a biochemical process which activates the transmission of a signal along the dendrite to the cell body. The signals from many dendrites are cumulated in the cell body and if the combined signal is strong enough it causes the neuron to fire producing an action potential as an output signal in the axon. The interplay between electrical transmission of information in the neuron and chemical transmission between cells is the basis for neural information processing [1].

The biological neuron can be modelled as an artificial neuron on the basis of four elements i.e. the synapses, dendrites, cell body and axon. In Fig.1.2b an information scheme is shown which reflects the main feature of the biological neuron. It can be concluded to be a mapping of many input stimuli into one output response.

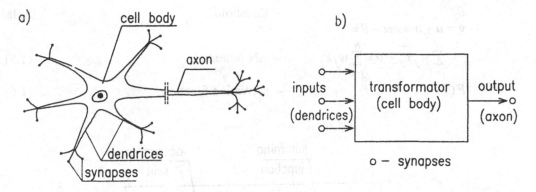

Fig.1.2: Biological and artificial neurons

The neurons are joined into neural networks. The extraordinary strength of neural network results from the immense number of neurons in the network and the number of connections between neurons. Let us only mention that the number of the human brain neurons is of order 10^{11} and the number of connections is of order 10^{15}. Typically, biological neurons are five to six orders of magnitude slower than silicon logic gates but massively parallel structure of neural networks makes them an extremely powerful computer – in many cases much more efficient than standard von Neumann type computers are.

Artificial neural networks (ANNs) are primitive models of the biological central nervous system. ANN models try to simulate very loosely the behaviour of the human brain. Especially ANNs are used for computations in the sense of the following triad [1]:

$$computation = storage + transmission + processing \,. \tag{1.1}$$

The use of ANNs for computation is called in short *neurocomputing*.

At the end of this Point it should be emphasised that in literature on ANNs there are many terms taken directly from biological terminology. For instance, *neurons* are synonyms to „computing units" and *neural networks* are understood as „artificial neural networks".

1.1.2. Model of artificial neuron

In Fig.1.3 an extended scheme of an artificial neuron (AN) is shown. The model has N inputs and one output. The neuron body is composed of two boxes: summing junction Σ and activation (threshold) unit F.

In the AN model shown in Fig.1.3 the following variables and parameters are used:

$$x = \{x_1, \ldots, x_N\} \qquad - \text{ input vector,} \tag{1.2}$$
$$w = \{w_1, \ldots, w_N\} \qquad - \text{ vector of weights,} \tag{1.3}$$
$$b = -\theta = w_0 \qquad\qquad - \text{ constant component (bias),} \tag{1.4}$$

$$\theta \qquad \text{– threshold,} \tag{1.4a}$$

$$v = u + b = net - \theta =$$

$$= \sum_{j=1}^{N} w_j x_j - \theta = \sum_{j=0}^{N} w_j x_j \qquad \text{– AN potential,} \tag{1.5}$$

$$F(v) \qquad \text{– activation function.} \tag{1.6}$$

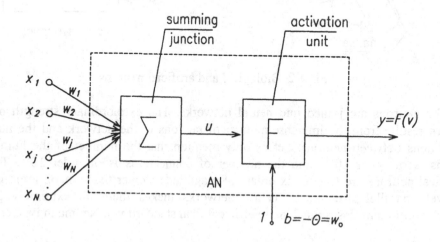

Fig.1.3: Model of an artificial neuron

In ANNs different activation functions are used. Below the most frequently used activation functions are listed:

a) linear function

$$F(v) = \beta v \ \text{for} \ \beta > 0 \ , \tag{1.7a}$$

which for $\beta = 0$ is also called the identity functions;

b) binary step function

$$F(v) = F(u - \theta) = \begin{cases} 1 & \text{for} \quad u > \theta, \\ 0 & \text{for} \quad u \le \theta; \end{cases} \tag{1.7b}$$

c) bipolar step function

$$F(v) = F(u - \theta) = \begin{cases} 1 & \text{for} \quad u > \theta, \\ -1 & \text{for} \quad u \le \theta; \end{cases} \tag{1.7c}$$

d) sigmoid function (logistic or binary sigmoid)

$$F(v) = \frac{1}{1 + \exp(-\sigma v)} \in (0,1) \quad \text{for} \quad \sigma > 0 ,$$

$$\frac{dF}{dv} \equiv F'(v) = \sigma F(1 - F) ; \qquad (1.7d)$$

e) bipolar sigmoid

$$F(v) = \frac{1 - \exp(-\sigma v)}{1 + \exp(-\sigma v)} \in (1,1) \quad \text{for} \quad \sigma > 0 ,$$

$$F'(v) = \frac{\sigma}{2}(1 + F)(1 - F) . \qquad (1.7e)$$

Graphics of the activation functions listed above are depicted in Fig.1.4.

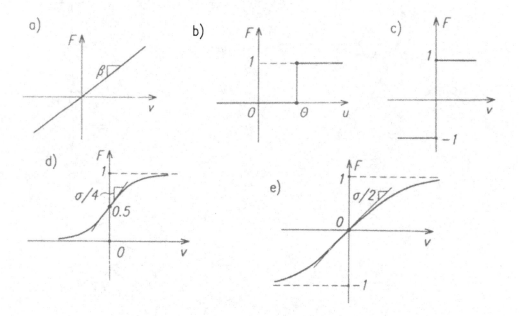

Fig.1.4: Activation functions: a) linear, b) binary step, c) bipolar step,
d) binary sigmoid, e) bipolar sigmoid

The binary sigmoid function simply called *sigmoid* is the most frequently used activation function. This function has the continuous derivative, cf. (1.7d)$_2$, and in algorithms it is used for mapping $x \rightarrow y = F(x) \in (0,1)$ instead of binary step function (1.7a).

In literature and practical applications various modifications and combinations of the above mentioned functions are used, e.g. piece-wise linear function, signum function, hyperbolic tangent function, etc. [2].

1.1.3. Types of neural networks

The structure and operation of a single neuron model are rather simple. A variety of complex connections of neurons makes them a powerful processing device, called *neural network* (NN). The arrangement of nodes and pattern of connection links between them is called the *architecture* of a neural network (type or structure of NN) [3].

There are three main NN architectures, cf.Fig.1.5:
1) feedforward NN,
2) recurrent NN,
3) cellular NN.

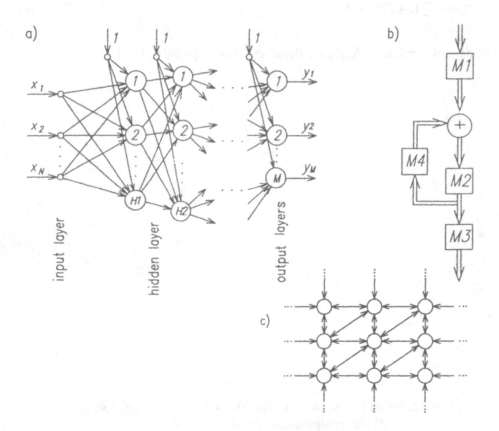

Fig.1.5: Three architectures of neural networks:
a) feedforward, b) recurrent, c) cellular

In the *feedforward neural network* signals are transmitted in one direction only, i.e. from inputs to outputs. The standard architecture of the feedforward NN corresponds to the layers of neurons. Usually neurons are not connected in the layer but they join the layer neuron with all the neurons of the previous and subsequent layers, respectively. In Fig.1.5a there are shown the input and output layers and the hidden layers between them. The structure of the feedforward NN can be written in short as:

$$N - H1 - H2 - ... - M, \tag{1.8}$$

where: N – number of inputs, M – number of output neurons, $H1$, $H2$,... – number of neurons in hidden layers.

Feedback and different direction of the signal transmission characterise the *recurrent neural network*, cf. Fig.1.5b. A topology of connections related to the lines of signals transmission is used in the *cellular neural network*, cf.Fig.1.5c.

1.1.4. Main features of neurocomputation

In this Chapter and in the book in general only computer simulations of neural networks are discussed. Basic information on NN hardware implementations can be found elsewhere, e.g. in [1].

Neural networks, applied to computations in the sense of triad (1.1), have special features. The first and most characteristic feature is natural, *massively parallel processing of information*. The next feature is *distributed processing and storing of information*. This implies a relatively low sensitivity of NN to its partial destruction, errors caused by noisy information, etc.

Instead of programming the *learning* (training) process is applied to NN. The learning is made by means of examples (patterns) taken from experimental tests or numerical computations by means of standard computers.

After a proper learning the neural network has the feature of *generalization*. This means that in the frame of considered problems the NN can efficiently process data other than those used in the learning process.

Another feature is the neural network *plasticity*. This corresponds to adaptivity of NN to new information.

The above mentioned and other features of neural networks are utilized in a great and increasing number of various applications. From the viewpoint of NN *applications in mechanics* the neurocomputing is especially efficient in the analysis of reverse problems, identification and assessment of material and structural properties. This means that neural networks can be treated as a *new tool* in the field of computational analysis.

Obviously, there are, problems which can be analysed perfectly by the existing computational methods and standard computers, e.g. direct problems or problems in which high numerical precision is desired.

That is why a prospective approach is to treat neural networks as a *complementary tool* for standard (von Neumann type) computers. This leeds to *hybrid strategies of computation* in which the neural procedures are incorporated into various computational algorithms and computer programs.

1.1.5. From the history of artificial neural networks

The mathematical modelling of neural networks was originated by W.Mc Culloch and W.Pitts in 1943. The research in this field has attracted attention of many scientists and engineers.

A great effort has been concentrated on the formulation of ANN fundamentals and their applications to pattern recognition, signal processing, statistical analysis and the basis of artificial intelligence. Among the main achievements the formulation and construction of the Perceptron by Rossenblat in 1958 should be mentioned, then the Madeline neural network by Widrow and Hoff in 1960.

The crucial point in the extensive development of ANN research and applications in science and technology was the publication of the book by Minsky and Papert in 1969. In the book the authors demonstrated such fundamental limits to the development of the perceptron-type neural networks that the enthusiasm and financial support for research on ANNs was significantly decreased for about ten years.

In spite of Minsky and Papert's prognosis the current leaders in the field continued their research on ANNs. Let us only mention the error backpropagation algorithm by Werbos, associative memory neural nets by Kohonen and Anderson, Hopfield's recurrent NNs, Grossberg and Carpenter's adaptive resonance theory.

A renaissance of artificial neural network started in the early 1980s. Publishing of book [4] by Rumelhart and McClelland in 1986 is treated as a crucial date. In 1987 the Institute of Electrical and Electronics Engineers (IEEE) originated a wave of conferences and congresses on neural networks. The rapidly increasing number of books and new periodicals reflect an extraordinary development of research in the field of ANNs. More information can be found in quoted books [1,2,3].

Let us return to the common acronym of artificial neural networks as 'neural networks' which are understood as a new tool to carry out the 'neurocomputing'. In such a sense NNs have been applied to the analysis of mechanics problems since late 1980s. Besides early papers presented at various conferences, the first paper published in an archival journal seems to be the one by Adeli and Yeh [5]. The first European conference with neural networks in the name was arranged in 1993, cf. [6]. In the years 1992-94 neural networks were used in mechanics problems developed in project [7].

During the last eight years an enormous increase of interest in neurocomputing can be observed. This involves also ANN applications in mechanics of structures and materials, cf. [8,9,10]. ANNs are used not only as a new tool for the analysis of simulation, identification and assessment problems, cf.[11], but neural procedures can be incorporated in hybrid NN/computational computer programs [12]. This approach opens the door to a new generation of computer simulations in which the best from among standard computer programs and ideas of neurocomputing create new, very promising prospects.

1.2. FEEDFORWARD NEURAL NETWORKS

1.2.1. Introduction to supervised learning

Let us discuss a single layer neural network with the binary step activation function, Fig.1.6a. This NN was called *Perceptron*. A single Perceptron neuron is shown in Fig.1.6b. The weight w_{ij} and biases (thresholds) $b_i = -\theta_i$ can be called *neural network parameters*.

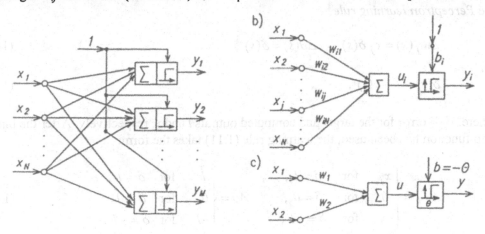

Fig. 1.6: a) Perceptron neural network, b) A single Perceptron neuron processing model, c) Two input Perceptron neuron

In Fig.1.6a a simple Perceptron neuron with two inputs only is shown. In order to simplify the notation the first subscript i, corresponding to the neuron number, is omitted. A single neuron can only categorize inputs x_i and x_2 into two classes, i.e. Class 0 corresponds to $y = 0$ and class for $y = 1$. For any values of w_1, w_2 and θ the line l can be computed from the equation

$$v \equiv w_1 x_1 + w_2 x_2 - \theta = 0 . \tag{1.9}$$

If line l separates Classes 0 and 1 the inputs x_1 and x_2 are *linearly separable*. As can be seen in Fig.1.7 for the functions OR and AND. In case of the function exclusive OR (XOR) the inputs x_1 and x_2 are not linearly separable since Class 1 is to be bounded by a curve c, cf. Fig.1.7.

Table 1.1: Functions OR, AND and XOR

Inputs		Output y		
x_1	x_2	OR	AND	XOR
0	0	0	0	0
0	1	1	0	1
1	0	1	0	1
1	1	1	1	0

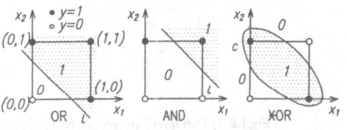

Fig. 1.7: Geometric representation of OR, AND and XOR functions

Parameters w_j and θ are computed in an iterative way

$$w_j(s+1) = w_j(s) + \Delta w_j(s) \ , \quad \theta(s+1) = \theta(s) + \Delta\theta(s) \ , \tag{1.10}$$

where: s – number of iterative step. Increments $\Delta w_j(s)$, $\theta(s)$ are computed by means of the *Perceptron learning rule*

$$\Delta w_j(s) = x_j \, \delta(s) \ , \quad \Delta\theta(s) = \delta(s) \ , \tag{1.11}$$

for

$$\delta(s) = t - y(s) \ , \tag{1.12}$$

where: δ — error for the target and computed outputs t and y, respectively. After the binary step function has been used, the learning rule (1.11) takes the form:

$$\Delta w_j = \begin{cases} x_j & \text{for} \quad \delta = 1 \ , \\ 0 & \text{for} \quad \delta = 0 \ , \\ -x_j & \text{for} \quad \delta = -1 \ , \end{cases} \qquad \Delta\theta = \begin{cases} 1 & \text{for} \quad \delta = 1 \ , \\ 0 & \text{for} \quad \delta = 0 \ , \\ -1 & \text{for} \quad \delta = -1 \ . \end{cases} \tag{1.13}$$

The application of learning rule (1.13) to the computation of parameters w_1, w_2, θ of the single neuron network initiates the iteration process which is convergent after a finite number of steps S if the learning patterns are linearly separable (Rosenbllat's theorem). This refers to functions OR and AND, cf. Fig.1.7.

In case of functions XOR one layer Perceptron network does not work and two layer neural network has to be applied. In Fig.1.8a a scheme of an appropriate NN is shown, where three Perceptron neurons are used. In Fig.1.8b the regions of inputs separation are shown by means of potentials $v_1 = w_{11}x_1 + w_{12}x_2 + b_1$ and $v_2 = w_{21}x_1 + w_{22}x_2 + b_2$ of neurons of the first layer.

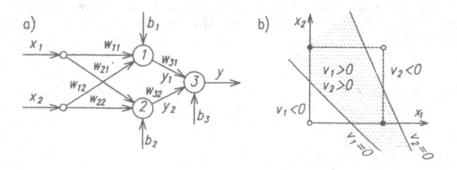

Fig.1.8: a) Two layer Perceptron NN for mapping XOR function,
b) Regions of input separation

Let us assume that *a smooth activation function F(v)* is introduced instead of the step function in the single layer network in Fig.1.6a. The network maps the input vector $x \in R^N$ into the output vector $y \in R^M$. Parameters w_{ij} and b_i of the number $M * (N + 1)$ are to be computed by means of a learning rule.

One of the learning rules corresponds to the *gradient method of the steepest descend.* According to this method the *Least Mean Squared Error* of the neural network is defined

$$E = \frac{1}{2} \sum_{i=1}^{M} (t_i - y_i)^2 , \qquad (1.14)$$

where: t_i, y_i – target and computed values of the *i*-th output, respectively. The increment of weights is computed according to the gradient formula:

$$\Delta w_{ij} = -\eta \frac{\partial E}{\partial w_{ij}} = -\eta \frac{\partial}{\partial w_{ij}} \left(\frac{1}{2} \sum_{i=1}^{M} (t_i - F(v_i))^2 \right) =$$

$$= \eta (t_i - y_i) \frac{dF}{dv_i} \frac{\partial}{\partial w_{ij}} \left(\sum_{k=1}^{N} w_{ik} x_k + b_i \right) = \qquad (1.15)$$

$$= \eta (t_i - y_i) \frac{dF_i}{dv_i} x_j = \eta \delta_i F_i' x_j .$$

where: η – *learning rate.* The increment of bias Δb_i can be deduced from (1.15) if we assume $x_0 = 1$ and $w_{i0} = b_i$. This leads to the following formula:

$$\Delta b_i = \eta \delta_i F_i' . \qquad (1.15a)$$

In case of identity function $F(v_i) = v_i$, derivative $F_i' = 1$ and *delta learning rule* is defined [*] :

$$\Delta w_{ij} = \eta \delta_i x_j , \qquad \Delta b_i = \eta \delta_i . \qquad (1.16)$$

This learning rule (Widrow and Hoff, 1960) can be treated as a special case of the *generalized delta rule* if formula (1.15) is written in the form:

$$\Delta w_{ij} = \eta \tilde{\delta}_i x_j , \quad \text{for} \quad i = 1,..., M; j = 0,1,...,N , \qquad (1.17)$$

where:

$$\tilde{\delta}_i = \delta_i F_i' , \qquad \delta_i = t_i - y_i , \qquad F_i' = dF_i / dv_i . \qquad (1.18)$$

---

[*] In what follows bias b_i is treated as weight w_{i0} corresponding to unite signal $x_0 = 1$.

The iteration process of the weights computation, shown in Fig.1.9, is called *supervised learning* (or learning with a teacher). This process is also called *training* of the neural network. The training set P is carried out on the basis of a *set of training patterns* (examples) which is composed of P pairs of known input and output vectors:

$$\mathcal{P} = \left\{ \left(x^{(p)}, \ t^{(p)} \right) \middle| \ p = 1, \ldots, P \right\}. \tag{1.19}$$

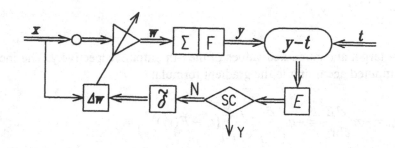

Fig.1.9: Scheme of supervised learning

The training process is composed of a sequence of forward and backward computations. One presentation of a training pattern p and updating of NN weights is called a *cycle*. One complete presentation of the entire training set during the learning process is called an *epoch*.

For a given training set, the backpropagation learning may proceed in one of two basic modes:

i) In the *pattern mode* the weight updating is performed after the presentation of *each* training example. This mode can be called *on-line updating*.

ii) In the *batch mode* the weight updating is performed after the presentation of *all* the training examples. The weight increments are computed *off-line*, usually by means of the average squared error:

$$E_{av} = \frac{1}{2P} \sum_{p=1}^{P} \sum_{i=1}^{N} \delta_i^2 \quad \text{where} \quad \delta_i = t_i - y_i, \tag{1.20}$$

$$\Delta w_{ij} = -\eta \ \frac{\partial E_{av}}{\partial w_{ij}} = -\frac{\eta}{P} \sum_{p=1}^{P} \delta_i^{(p)} \ \frac{\partial \delta_i^{(p)}}{\partial w_{ij}^{(p)}}. \tag{1.21}$$

1.2.2. Backpropagation algorithm

The learning method associated with the gradient method of steepest descent can usually be applied to the updating of multilayer neural networks of the structure shown in Fig.1.5a.

Without loosing the idea of generality we can discuss a two layer network. Let us consider the k-th neuron of the hidden layer. In Figs 1.10a,b the feedforward transmission of signals and backpropagation weight updating are shown.

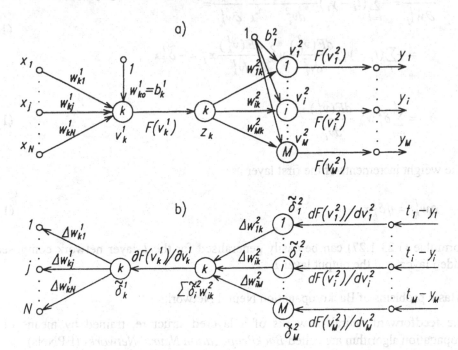

Fig. 1.10: a) Feedforward transition of signals, b) Error backpropagation

The neural network LSM error is:

$$E = \frac{1}{2}\sum_{i=1}^{M}(t_i - y_i)^2 = \frac{1}{2}\sum_i \left[t_i - F(v_i^2) \right]^2 = \frac{1}{2}\sum_i \left[t_i - F\left(\sum_{k=0}^{H} w_{ik}^2 z_k \right) \right]^2 =$$

$$= \frac{1}{2}\sum_i \left[t_i - F\left(\sum_k w_{ik}^2 F\left(\sum_{j=0}^{N} w_{kj}^1 x_j \right) \right) \right]^2 ,$$

(1.22)

and the derivative:

$$\frac{\partial E}{\partial w_{ik}^2} = -(t_i - y_i)\frac{dF(v_i^2)}{dv_i^2} z_k = -\tilde{\delta}_i^2 z_k .$$

(1.23)

It enables us to compute the weight increments of the output layer:

$$\Delta w_{ik}^2 = \eta \tilde{\delta}_i^2 z_k .$$

(1.24)

A similar proceeding is related to the first (hidden) layer:

$$\frac{\partial E}{\partial w_{kj}^l} = -\sum_{i=1}^{M}(t_i - y_i)\frac{dF(v_i^2)}{dv_i^2}\frac{\partial v_i^2}{\partial z_k}\frac{\partial z_k}{\partial w_{kj}^l} =$$

$$= -\sum_{i}(t_i - y_i)\frac{dF(v_i^2)}{dv_i^2}w_{ik}^2\frac{dF(v_k^l)}{dv_k^l}x_j = -\tilde{\delta}_k^l x_j,$$

(1.25)

where:

$$\tilde{\delta}_k^l = \sum_{i=1}^{M}\tilde{\delta}_i^2 w_{ik}^2\frac{dF(v_i^l)}{dv_i^l}.$$

(1.26)

The weight increments of the first layer are:

$$\Delta w_{kj}^l = \eta\tilde{\delta}_k^l x_j.$$

(1.27)

Formulae (1.23-1.27) can be easily generalised for the *L*-layer network composed of *L-1* hidden layers and the output layer *L*.

1.2.3. Basic problems of Backpropagation Neural Networks

The feedforward neural networks of a layered structure, trained by means of the backpropagation algorithm are called *BackPropagation Neural Networks* (BPNNs).

From among many questions concerning the formulation and applications of BPNNs the basic questions are:

1. *Data analysis* associated with the underlying problems (selection of training and testing patterns, preprocessing of data etc.);

2. *Architecture* (arrangement of layers, number of neurons and connections between them), *network parameters* (synaptic weights) and *learning parameters* of BPNNs;

3. *Learning* of the neural networks and related questions (selection of activation functions, method and parameters of learning, initial values of weights, stopping criteria etc.);

4. *Generalization* of the neural network, i.e. the ability of BPNN to produce reasonable responses to input patterns of the problem analysed, which are not identical with training patterns.

The questions listed above have to be answered since they correspond to the formulation, computational procedures and operation advantages of BPNNs. But in fact these questions are interconnected and they have to be considered paralelly or in a different order.

1.2.3.1. Mechanical problems and data arrangement

In general data can be produced by means of theoretical models or taken from experiments on material models. Such data can be used for both training and testing of neural nets.

In Table 1.2 the input and output data correspond to the problems considered and input/output variables used in mechanical systems [11].

Table 1.2: Mechanical system problems and relevant data

Problems	Inputs	Outputs
1. Mechanical system (MS) response simulation	Excitation variables and system parameters	Response variables
2. MS excitation simulation (identification)	Response variables and system parameters	Excitation variables
3. MS parameter identification 4. MS excitation and/or response assessment	MS excitation and response variables	Parameters of MS
	All relevant measures of system conditions to be assessed	Assessment of system condition

In case of one-to-one correspondence the computed response or a part of response variables can be used for the simulation of excitation. The data taken from theoretical models can be used for the BPNN training and experimental data can serve for the testing. Of course, it is also possible to change them or mix various data for the network training and testing.

The assessment problems are usually a kind of classification problems. That means that in the space of input variables it is possible to distinguish *classes of data*. In Fig.1.11 two classes C_1 and C_2 are shown with a known assignment of outputs, e.g. the input variables marked by the cross × correspond to the output value 1 and the circles o correspond to the zero output value.

In Fig.1.11a classes C_1 and C_2 are linearly separable, and in Fig.1.11b a case of nonlinearly separable classes is shown. A more complicated case of nonlinearly separable data is shown in Fig.1.11c

In case of spread patterns they may be grouped into *clusters*. Such clusters, corresponding to classes C_1 and C_2 are shown in Fig.1.12.

Clustering of data can be very useful for the selection of training patterns and their validation by means of appropriate testing patterns.

The selection of input/output data strongly depends on the problem at hand. The principal and uncorrelated variables of space behaviour are preferable. From this viewpoint the *dimension analysis* is recommended and dimensionless variables should be used [13].

Preprocessing of data is commonly used in BPNNs. Scaling and normalization are used especially with respect to the output neurons. In case of binary sigmoid the range

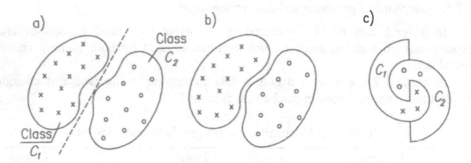

Fig.1.11: a) Linearly separable classes of data,
b) c) Nonlinearly separable classes

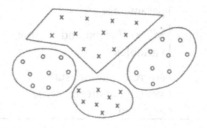

Fig.1.12: Grouping of patterns into clusters

[0.1, 0.9] is preferable instead of (0.0, 1.0). Vector classification and clustering of input data can be easier after their preprocessing.

Another possibility is *data transformation*. A good example is the damage assessment of structures subjected to earthquake excitations. Transformation from the time domain to the frequency domain and digitalization of continuous distribution of variables enable us to compute input data suitable in the damage analysis [14].

Multilevel BPNNs are sometimes used to preprocess data taken from various domains. Three level BPNNs were used in [15] to identify the relations between geometrical and design variables and characteristic parameters of semi-rigid steel connections.

An extension of the input information by means of variable products corresponds to the *functional link net* also called the Pao networks [16]. Such a net can have, in general, a lower number of synaptic weights and the learning process is faster [17].

In Fig.1.13a the Pao neural network is shown for XOR problem. Because of term $x_1 x_2$ only one neuron with binary step activation function and 5 network parameters are needed instead of 9 parameters in two-layered net shown in Fig.1.8. Term $x_1 x_2$ also causes nonlinear separability of variables x_1, x_2, as shown in Fig.1.13b.

Functional link NNs were also applied successfully to the analysis of plane trusses [18].

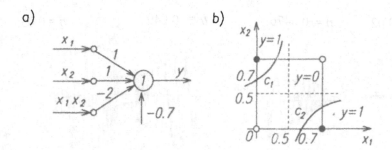

Fig. 1.13: a) Pao neural network and values of synaptic weights for XOR problem,
b) Nonlinear separation of variables x_1 and x_2

A very important question associated with BPNNs is the *number of patterns* and *splitting* them into the training and testing sets. These questions are strongly related to the architecture and generalization properties of BPNNs. That is why the above questions will be discussed in subsequent Points.

1.2.3.2. More about learning methods

If we assume that a set of patterns was well selected and the architecture of neural network was „optimally" formulated then we can to focus our attention on learning methods.

The *learning process* corresponds to computation of the network parameters, i.e. synaptic weights w_{ij} (biases are included as w_{i0} weights) at the fixed or adaptively computed learning parameters η or η_{ij}. An algorithm used in the learning process is called a *learning method*.

The network error function E constitutes the *errror surface* in the space of parameters w_{ij}. The NN learning methods are in fact methods of mathematical programming of searching a minimum of function E in the space of design variables w_{ij}. The learning rate η can be interpreted as a step parameter in the formula for the increment of design vector w :

$$\Delta w = \eta\, p , \tag{1.28}$$

where: p – vector of search direction.

In case of the steepest descent gradient method the directional vector is

$$p = -\nabla E , \quad \text{where} \quad \nabla E = \mathbf{grad}\ E = \partial E/\partial w_{ij} . \tag{1.29}$$

Formula (1.29) was used in (1.15) as the basis for the *Classical BackPropagation* learning method (CBP).

The convergence of CBP method depends on the shape of error surface and the value of learning rate η. In Fig.1.14, taken from [19], a simple case of quadratic surface is shown with minimum at the point +. Starting from an initial vector $w(0)$ at the point o subsequent increments at $\eta = $ const. correspond to points .

$\eta \doteq 0.02$ $\eta = 0.0476$ $\eta = 0.049$ $\eta = 0.055$

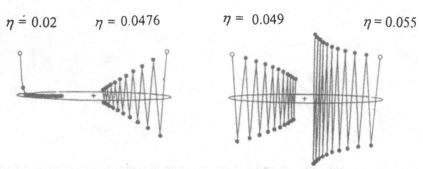

Fig.1.14: Gradient descent at a flat quadratic surface

At a small value of η (in [19] it was $\eta = 0.02$) the gradient descent shown at the left hand side in Fig.1.14 is along nearly the best line of search but because of the flatness of error surface a great amount of steps are needed to approach the minimum $+$. The next gradient descents ($\eta = 0.0476, 0.049, 0.055$) caused quicker approaching the minimum but oscillations appeared and at a higher value of $\eta = 0.055$ the iteration process was divergent.

The example discussed above points out two main features of CBP learning method: i) the method is usually slowly convergent (especially in the regions of the error surface E), ii) the rate of convergence increases at higher values of learning rate η (it takes place at the slopy region of surface E) but at higher values η the iteration process can be divergent.

Momentum term

The disadvantages of CBP mentioned above made us develop more efficient and stable learning methods. One of the improvements of CBP is adding the *momentum term* in formula (1.29)

$$\Delta w(s) = -\eta \nabla E(s) + \alpha \, \Delta w(s-1) , \qquad (1.30)$$

where: s – the number of iteration step. In case of the flat part of surface E increments Δw are nearly equal in subsequent steps, i.e. $\Delta w(s) \approx \Delta w(s-1)$ and formula (1.30) takes the form:

$$\Delta w(s) \approx -\frac{\eta}{1-\alpha} \nabla E(s) = -\tilde{\eta} \nabla E(s) . \qquad (1.31)$$

The above formula leads to the conclusion: at the flat region of error function the increased learning rate $\tilde{\eta} = \eta / (1 - \alpha)$ can significantly grow. Let us assume, for instance, the momentum parameter to be $\alpha = 0.9$; then the value of learning rate increases 10 times, i.e. $\tilde{\eta} = 10\eta$.

In many papers and books the values of the rate learning and momentum parameters are kept constant during the learning process. It is much more reasonable to apply an adaptive approach, i.e. to switch on the momentum term only at the flat regions of function E.

In the MATLAB NN Toolbox [20] the following formula is suggested

$$\Delta w_{ij}(s) = -(1-\alpha)\eta\, g_{ij}(s) + \alpha\, \Delta w_{ij}(s-1) , \qquad (1.32)$$

where: $g_{ij} = \partial E/\partial w_{ij}$ and $\alpha \neq 0$ if $E(s)/E(s-1) < 1.04$. In Fig.1.15 two graphics from Demo BP5 example [20] are shown. In Fig.1.15a the gradient descent is shown corresponding to CBP learning method and in Fig.1.15b the operation of momentum term can be observed. Due to switching on the momentum term an increase of the learning rate $\tilde{\eta}$ causes a „jump out" of a local minimum, cf. Fig.1.15b, and reaching a better solution corresponding to a lower value of the sum squared error (cf. Fig.1.15c)

$$SSE = \sum_{p=1}^{L}\left(t^{(p)} - y^{(p)}\right)^{2} . \qquad (1.33)$$

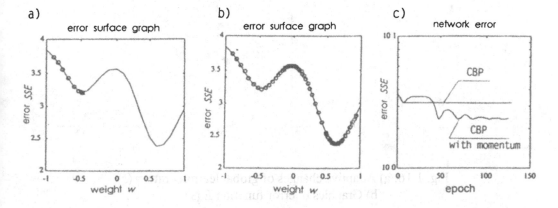

Fig.1.15: a) Approaching a local minimum for classical BP algorithm, b) Operation of CBP and momentum, c) Learning process for CBP and CBP with momentum

Fast learning methods

More refined methods of mathematical programming were used as „*fast" learning methods*. From among them the conjugate gradient method and second-order methods (Newton and Pseudo-Newton methods) are worth mentioning [1]. Those methods need many more arithmetic operations than CBP with momentum. The fast methods mentioned are not general since their efficiency strongly depends on the problem at hand.

A special group of fast methods are *adaptive methods*. They can be split into *global adaptive methods* and *local adaptive methods*.

The *global learning rate* η can be changed according to an adaptive criterion. In MATLAB [20] the following adaptive formula is given:

$$\eta(s) = \begin{cases} \eta^- \, \eta(s-1) & \text{if} \quad SSE(s) > er \cdot SSE(s-1) \ , \\ \eta^+ \, \eta(s-1) & \text{if} \quad SSE(s) < er \cdot SSE(s-1) \ , \end{cases} \tag{1.34}$$

where: $\eta^- = 0.7$, $\eta^+ = 1.05$, $er = 1.04$. In Fig.1.16 there is shown an influence of the adaptation according to formula (1.34) on the global learning rate $\eta(s)$ and the corresponding error function $E(s)$, cf. [17].

In local adaptive algorithms a *local learning rate* η_{ij} is computed for each weight:

$$\Delta w_{ij} = -\eta_{ij} g_{ij}, \tag{1.35}$$

where: $g_{ij} = \partial E / \partial w_{ij}$. The local learning methods are related either to the first order gradient algorithms (Almeida's, Delta-Bar-Delta, Rprop methods) or to the second order gradient algorithms (QuickProp, QRprop methods).

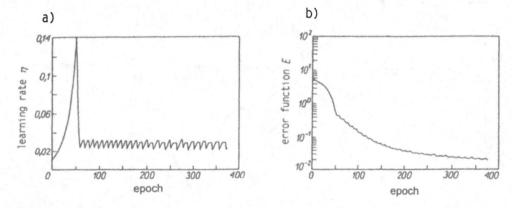

Fig. 1.16: a) Adaptive changes of global learning rate $\eta(s)$,
b) Graphics of error function $E(s)$

The author and his associates had a good experience with the Rprop (Resilient-propagation) method so only this learning method is discussed below.

The weights are updated according to the following Rprop formula, cf.[1]:

$$\Delta w_{ij}(s) = -\eta_{ij}(s) \operatorname{sgn} g_{ij}(s) \ , \tag{1.36}$$

where:

$$\eta_{ij}(s) = \begin{cases} \min\left(\eta^{+}\eta_{ij}(s-1),\ \eta_{max}\right) & \text{for} & g_{ij}(s)g_{ij}(s-1) > 0, \\ \max\left(\eta^{-}\eta_{ij}(s-1),\ \eta_{min}\right) & \text{for} & g_{ij}(s)g_{ij}(s-1) < 0, \\ \eta_{ij}(s-1) & \text{otherwise.} \end{cases} \qquad (1.37)$$

According to [21] the fixed parameters used in (1.37) are: $\eta^{+} = 1,2$, $\eta^{-} = 0.5$, $\eta_{max} = 50$, $\eta_{min} = 10^{-6}$.

The Rprop algorithm is of heuristic type and it works surprisingly well in the flat regions of the error function.

More information about fast learning methods can be found in [1,17].

Weights initiation and global minimum

All the above mentioned methods have a basic disadvantage of gradient algorithms, which is related to reaching *a local minimum* of error function. The placement of the minimum can strongly depend on *initial values of weights $w_{ij}(0)$*.

Usually the initial weights are randomly selected from the range [-a,a]. In literature there are many suggestions how to estimate the value of parameter a, cf. [2,17]. For instance Nguyen and Widraw propose the evaluation $a = (H)^{1/n_{in}}$ for hidden layers, where: H — number of neurons in the layer, n_{in} — number of inputs to the considered neuron. For the output layer a is 0.5.

From among other methods of weight initiation let us mention only the application of Hebbian learning and self-organizing algorithms [2,17].

The main problem of 'how to reach *the global minimum* of error function?' can be solved experimentally starting from different initial weights. It is more reasonable to apply methods of *soft optimizations*, e.g. *simulated annealing* or *genetic algorithms* [1]. Chapter 2 of this book is devoted to the application of genetic algorithms to the computation of weight values.

Stopping criteria

The iteration process is ended by a *stopping criterion* (SC in Fig.1.9). Various stopping criteria can be found in literature, cf. [2]. From among them two criteria of practical value are commonly used. They are related to the number of epoch presentations in the following way:

1) such number of epochs s_* that $E(s_*) < \varepsilon$,

    where: ε – prescribed error for the network error function E;   (1.38.1)

2) prescribed number S with the corresponding network error $E(S)$.   (1.38.2)

Other very practical stopping criteria can be formulated on the basis of learning and testing processes. This question is discussed in Point 1.2.3.3.

Obviously, instead of the LMSE error E other measures of error function can be used. The most frequently used error measures are listed below:

a) Least Mean Squared Error and average LMSE:

$$E = \frac{1}{2} \sum_{p=1}^{P} \sum_{i=1}^{M} \left(t_i^{(p)} - y_i^{(p)} \right)^2 \; , \; E_{av} = \frac{E}{P} \; , \tag{1.39}$$

where: p – the number of pattern, i – the number of output neuron.

b) Sum Squared Error, cf. (1.33):

$$SCE = 2E \; , \tag{1.40}$$

c) Mean Squared Error:

$$MSE = \frac{SSE}{P} \; , \tag{1.41}$$

d) Root Mean Squared Error:

$$RMSE = \sqrt{MSE} \; . \tag{1.42}$$

The above listed errors can be easily computed at the end of each epoch or they can be used as a network error introduced to the BP algorithm. Of course, in such an error application the BP formulae (1.22-1.27) should be formulated adequately.

1.2.3.3. Generalization, cross-validation and testing.

Let us assume that in the behaviour space X we can separate a subspace $\mathcal{R}$ in which the input and output variables obey a set of rules (relations). From subspace $\mathcal{R}$ it is possible to select a set of representative patterns $\mathcal{P}$ which is split into a learning and testing sets, L and T, respectively. Additionally, a validation set $\mathcal{V}$ can be selected from learning set L, cf. Fig.1.17.

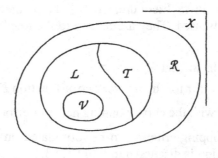

Fig.1.17: Different sets of patterns in behaviour space X

Network generalization

The neural network is learned on representatives of space $\mathcal{R}$. This means that the main goal of the learning process is to memorise by the network the rules which constitute the relations between the input and output data in $\mathcal{R}$. The learnt network should have *generalization properties*, i.e. the network trained on the learning set L should well operate on all patterns from $\mathcal{R}$. In order to verify the generalization of the network it is tested on a representative set of testing patterns $\mathcal{T}$.

The generalization properties of BPNNs are influenced by three factors: 1) the size and efficiency of the learning set, 2) the architecture of the network, and 3) the physical complexity of the problem at hand. Clearly, we have no control of the problem considered. The learning set is commonly determined so usually we are interested in the following problem, cf. [2]:

- The size of the learning set is fixed, and the issue of interest determines the best architecture of the network to achieve good generalization.

The problem is illustrated in Fig.1.18, on examples Demo BP4, LM2 and LM3 taken from [20]. The approximation of a function was carried out with respect to 21 target points. If a hidden layer has only $H = 2$ neurons then the network cannot be well trained, i.e. the learning error is significant, and the testing errors for intermediate points are also high, cf.Fig.1.18a. The network with $H = 5$ neurons has a very small learning error and good generalization (interpolation properties), cf. Fig.1.18b. If the number of neurons in the hidden layer increases up to $H = 40$ then the network is *overfitted*. It is shown in Fig.18c that an overfitted network is very well trained but its generalization (interpolation of the function) is poor.

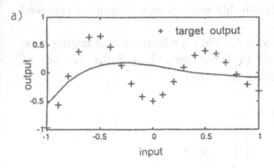

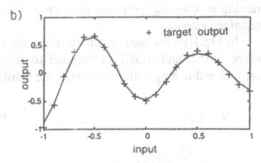

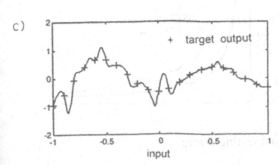

Fig. 1.18: Approximation of a function for $L = 21$ learning patterns by means of BPNN of structure *1-H-1* at different number of neurons in the hidden layer:

The relation between the number of neurons (number of synaptic weights) and the number of learning patterns is not straightforward. This question is discussed in Point 1.2.3.4.

Validation and testing

Usually we choose the „best" network by estimation of the network architecture and parameters within a set of candidate networks. In this context a standard tool is *cross-validation* [2]. A validation subject V is selected from learning set L (typically 10 to 20 percent of L). Then the candidate networks are trained on subset $L' = L \backslash V$ and are *validated* by means of set V.

The motivation of validation is to assess the performance of various candidate models as networks, and thereby choose the „best" one. After validation the generalization properties of the resulting network are verified by a testing set T.

An interesting approach to cross-validation is suggested in [22]. In case of small training sets it is possible to change subset V and even reduce it to a single training pattern. In this way „noisy" and „clean" data can be separated. This strategy was used in [23] for the evaluation of experimental data.

Quite often the cross-validation stage is omitted and the estimations described above are performed by the training set. In this way the resulting network can have good generalization performance.

Renewed stopping criteria

A number of computer simulators of neural networks offer an option to couple the learning and testing procedures [21]. This means that after each training epoch both the training and testing errors are computed. This possibility enables us to formulate *renewed stopping criteria*,

In Fig.1.19 the *MSE* error functions are shown for both the learning and testing sets, where *MSEL* and *MSET* correspond to formula (1.41) if instead of P the number of learning and testing patterns, L and T are substituted, respectively.

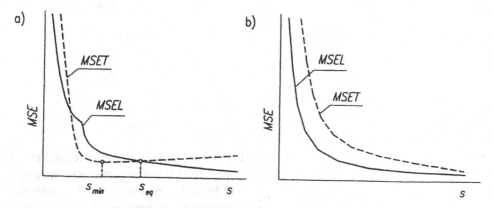

Fig. 1.19: Learning and testing errors

Two possible stopping criteria are associated with Fig.1.19a:

1) such number of epochs s_{min} that *MSET* is minimal, (1.43.1)

2) such number of epochs s_{eq} that $MSET \approx MSEL$. (1.43.2)

From the above listed criteria especially the second one is of practical value since it enables us to reach an „optimal" network. In case shown in Fig.1.19b we have to use stopping criteria of type (1.38).

1.2.3.4. Multilayered NN architectures

At the beginning of this Point let us discuss one use of neural networks as *universal approximators*. The use concerns the approximation of a continuous mapping *f*. The theoretical background was taken from *Kolmogorov's mapping existence theorem* (Kolmogorov, 1957, Sprecher, 1965) and finally it was expressed in the terminology of neural nets by Hecht-Nielsen, (1987) [2]:

- Given any continuous function $f: I^N \rightarrow R^M$, where *I* is the closed unit interval [0,1], *f* can be represented exactly by a feedforward neural network having *N* input neurons, $2N + 1$ neurons in one hidden layer, and *M* outputs.

In case of a discontinuous function the number of hidden layers has to be bigger than one.

A straightforward application of Kolmogorov's theorem to design a neural net is far from simple. From among many factors the selection of activation functions should be considered. That is why practical realizations of „optimal" neural networks can have significantly different number of layers and neurons in them than can be deduced from Kolmogorov's theorem.

In what follows we return to *computable sets of patterns* and we continue the analysis of NN architecture related to the number of layers, the number of neurons in the layers, and the connections between neurons. The arrangement of an architecture should correspond to the problem at hand and selected sets of patterns. The arrangement obviously depend on the assumed activation functions.

Number of training patterns

A theoretical background to the NN architecture design is associated with the *generalization problem*. At the fixed type of activation function the question is: What is the relation of NN architecture and number of training patterns.

A measure for generalization is difficult to be defined. The main parameter usually applied to generalization estimation is the *Vapnik-Chervonenkis dimension VCdim*. Parameter *VCdim* is the maximum number of training examples that can be learned for all possible binary labelings of classification functions (usually binary step functions). It is clear that *VCdim* is closely related to the separating capacity of a surface encountered in pattern classification [2,17].

The number of representative learning patterns *L* should be bigger than *VCdim*. Unfortunately, for bigger neural networks estimation of *VCdim* is very difficult. For instance, in case of NN with one hidden layer and discontinuous activation the following evaluation is valid [17]:

$$2 \, \text{enter} \left[\frac{H}{2}\right] N \le VC \, dim \le 2N_w \left(1 + \log N_N\right), \tag{1.44}$$

where: N – number of inputs, N_w – number of network synaptic weights, N_N – total number of neurons, H – number of neurons in the hidden layer. The lower bound equals approximately the number of weights connecting the input and hidden layers. The upper bound, however, is more than twice the total number of the net weights. That is why the estimation $VCdim \approx N_w$ is assumed for networks with an arbitrary number of hidden layers.

Knowing the value of parameter $VCdim$ the number of training patterns can be evaluated. As expressed above, the following relation should be fulfilled: $P > VCdim$.

In [2] a different estimation is shown for one hidden layer and bipolar activation functions, as well as for the bipolar input and output training values x_i, $t_i \in [-1, 1]$:

$$L \ge \frac{32N_w}{\varepsilon} \ln\left(\frac{32H}{\varepsilon}\right) \approx \frac{N_w}{\varepsilon}, \tag{1.45}$$

where: ε – fraction of errors permitted on test. Thus, with an error of 10%, the number of training patterns L should be approximately ten times the number of synaptic weights in the network.

Heuristic approaches to NN architectures design

The above mentioned estimation provides a kind of worse-case formulae. In practice there can be a huge numerical gap between „optimal" experimentally designed NN architecture (number of synaptic weights) and theoretical predictions.

In practice heuristic-type estimations are prefered to start with a first architecture and then improve it. Usually one-hidden layered network with sigmoid activation functions is applied. Practically we do not introduce more than two hidden layers. This reflects general features of those layers.

The first hidden layer extracts *local features*. Some neurons in this layer are used to partition the input space into regions, and other neurons learn the local features of those regions.

The second hidden layer extracts *global features*. A neuron in the second layer combines the outputs of neurons in the first hidden layer operating on a particular region of the input space, and thereby learns the global features for that region.

There are two strategies in experimentally designed architecture:

1) Start from a very simple architecture (one hidden layer, e.g. $H \approx \sqrt{NM}$ where: H, N, M – number of neurons in the hidden, input and output layers, or even $H = 1$ or 2) and check the generalization by means of parallel training and testing processes. In case

when the training error is high the complexity (number of neurons in the hidden layer) of network should increase.

2) Start from a more complex architecture (two hidden layers, e.g. $H1 \approx N$, $H2 \approx M$). In case of overfitting the network is simplified, i.e. the number of neurons $H1$ and $H2$ is decreased. As in strategy 1) the parallel training and testing processes enable us to estimate the generalization properties of considered networks.

There are many experimental suggestions about the architecture parameters, cf. e.g. [24]. Unfortunately, the majority of such recommendations is strongly problem - dependent so good advice is: be careful with their generalization.

Other problems of architecture design

The two mentioned strategies correspond to two kinds of algorithms which are formulated for automatic computation of optimal networks. *Cascade algorithms* start from simple to complex networks (the network-growing approach). *Pruning algorithms* eliminate non-active weights. References on those approaches can be found in [1,2,17,19].

So far only layered neural networks have been discussed. Additional effects can be achieved if *over-layer connections* are introduced. This is shown in Fig.1.20 on a network for the XOR problem. Due to direct connections of the input points with the output neuron (over the hidden layer) the total number of neurons is two and seven weights are needed. In case of a standard two-layered network the corresponding numbers are: 3 neurons and 9 weights, cf. Fig. 1.8a.

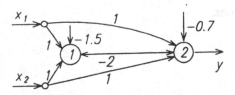

Fig.1.20: Network for the analysis of XOR problem with over-layer connections

1.2.4. RBF networks

A special type of neuron network is associated with *Radial Basis Function* (RBF) $\varphi(\|x - c\|)$, where c is the position vector of the *centre*. The application of RBFs corresponds to clustering of data and in case of regions of input variable condesation this approach can be more efficient than the application of sigmoid type functions.

In Fig.1.21a *RBF neural network* is shown. The network is composed of two layers, i.e. hidden and output layers, respectively (as in multilayer networks the input layer is not counted). The RBF neurons are associated with functions φ_k, where $k = 1,...,H$ and H is the number of RBF neurons. Without loosing the idea of generality only one output neuron is considered.

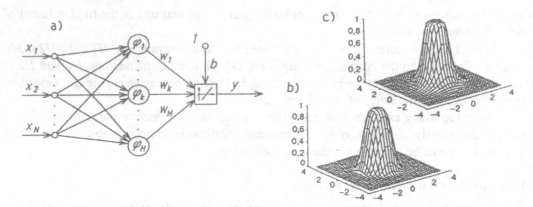

Fig.1.21:a) Classical RBF network, b) Gaussian RB function $\varphi(r) = \exp(-r^2 / 2\sigma^2)$,

c) Hardy's RBF $\varphi(r) = 1 / (r^2 + \sigma^2)$

The network shown in Fig.1.21a can be called a *classical RBF network*. This network has a linear output neuron corresponding to the function:

$$y = \sum_{k=1}^{H} w_k \varphi_k + b ,$$ (1.46)

where: φ_k are radial basis functions:

$$\varphi_k = G\left(\left\| x - c^k \right\|_Q \right) .$$ (1.47)

In (1.47) the weighted Euclidean norm with matrix Q is defined:

$$\| v \| = \sqrt{\| v \|_Q^2} , \quad \| v \|_Q^2 = (Qv)^T (Qv) = v^T Q^T Q v = v^T C v .$$ (1.48)

Matrix Q enables us to transform (scale) vector $v \in \mathcal{R}^N$ with respect to different axes j', i.e. $v'_j = Q_j v$ for $j = 1,...,N$.

The most commonly used RBF is associated with Gaussian functions:

$$G_k = \exp \sum_{j=1}^{N} \left(-\frac{(x_j - c_j^k)^2}{(2\sigma^k)^2} \right) = \exp\left(-[x - c_k]^T D [x - c_k] \right) ,$$ (1.49)

where: D – diagonal weighting matrix $D = \{D_{kk}\} = \left\{ 1 / (2\sigma_k^2) \right\}$.

In Fig.1.21b Gaussian function was depicted for $x = \{x_j\}$, $r = \|x - c\|$, $c = 0$, $\sigma^2 = 2$ for $j = 1,2$. From among other possible RBFs the Hardy function is shown in Fig.1.21c.

An application of the classical RBF network is shown to the solution of XOR problem [2]. The network with two Gaussian functions of the form (1.49) was used to approximate the function:

$$y\left(x^{(p)}\right) \approx t^{(p)}, \tag{1.50}$$

where: $x^{(p)}, t^{(p)}$ – input vectors and output values corresponding to patterns p listed in Table 1.2 for XOR problem.

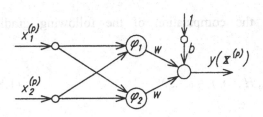

Fig.1.22: RBF network for XOR problem

Table 1.2: Input-output transformation by means of network in Fig.1.22

p	$x^{(p)}$	$t^{(p)}$	computed $y^{(p)}$
1	(0,0)	1	+0.901
2	(0,1)	0	-0.012
3	(1,0)	0	-0.012
4	(1,1)	1	+0.901

In this example two centres were used with vectors $c_1 = \{1,1\}$, $c_2 = \{0,0\}$, and variance $(\sigma^k)^2 = 2$. The problem was reduced to the solution of the following equation: $Gw = t$, where: $w = \{w,w,b\}$, $t = \{1,0,1,0\}$ and G is not a square matrix. A pseudoinverse matrix $G^+ = (G^T G)^{-1} G^T$ was computed in [2] and values $w = 2.284$, $b = -1.692$ were finally found. In Table 2 the computed outputs $y^{(p)}$ are given.

The classical RBF network can be generalised to the *Hyper Radial Basis Function* (HRBF) network. In the RBF network variance σ_k^2 was associated with each k-th neuron. In case of the HRBF network the scaling matrix $Q^k = \left[Q_{ij}^k\right]$ and variance $(\sigma_j^k)^2$ are associated with both the hidden neurons $k = 1,...,H$ and the components of the input vector x_j for $j = 1,...,N$.

The principal problems of RBF networks are:

1) selection of the number of RB functions (the number of neurons in the hidden layer H),
2) selection of centres and parameters of RBFs,
3) computation of network weights w_k for $k = 0,...,H$, where $w_0 = b$, cf. Fig.1.21a.

Let us assume that the number H is fixed. Then various strategies can be applied to the estimation of RBF centres and parameters as well as for the computation of network weights [2,17]. One of the strategies fully corresponds to supervised learning and especially the BP algorithm can be used.

The LMSE error (1.14) is related to RBF neurons:

$$E = \frac{1}{2} \sum_{k=0}^{H} \left[t - w_k \varphi_k(x) \right]^2 , \qquad (1.51)$$

where, let us assume, that $\varphi_k(x)$ are HRB Gaussian functions:

$$\varphi_k(x) = \exp\left(-\frac{1}{2} \left[Q^k \left(x - c^k \right) \right]^T \left[Q^k \left(x - c^k \right) \right] \right) . \qquad (1.52)$$

The classical BP algorithm needs the computation of the following gradient components:

$$\frac{\partial E}{\partial w_k}, \ \frac{\partial E}{\partial c_j^k}, \ \frac{\partial E}{\partial Q_{ij}^k} \quad \text{for} \quad k = 0,1,\dots,H, \quad i,j = 1,\dots,N. \qquad (1.53)$$

Obviously, parameters w_k, c_j^k, Q_{ij}^k can be computed in an iterative way using global or local learning rate parameters η_r^k, where r corresponds to the parameters in (1.53).

Other approaches are related to random selection of centres or to the application of self-organizing selection associated with unsupervised learning, cf. Section 1.3.

From among the methods which have been used for the estimation of the number of RBFs the LSM method combined with the standard Gram-Schmidt orthogonalization procedure is worth pointing out [2].

More about the problem mentioned above can be found in references given in the books [2,17].

The RBF neural network (RBFN) has many advantages in comparison with sigmoid type networks. RBFs are related to local approximations around centres in the space of input variables. This feature can well match the considered physical problems. In case of RBFNs it is easier to initiate the starting values and the learning process can be shorter than in sigmoid networks.

RBFNs have been successfully applied in classification problems, for the approximation of multivariable functions and prediction problems. In applications RBFNs were used for image processing and speech recognition for time-series analysis, adaptive equalization and medical diagnosis [2]. An interesting application of RBFN to prediction of parameters of composite structures was discussed in [25]. RBFNs are also used as regularization networks [2] which enable us to ommit the overfitting of neural approximation.

1.3. UNSUPERVISED LEARNING

The basic point of the previous Section was supervised learning, i.e. the desired output for each of the training vector is presented to the network.

Unsupervised learning is a means of modifying the NN weights without specifying the desired output for any input patterns. The network with unsupervised learning is a self-organizing network. This means that provision is made for a *task-independent measure* of the quality of representation that the network is required to learn, and the free parameters of the networks are optimised with respect to that measure. In such a way the tuned network has formed internal representations for encoding features of the presented input [1]. Unsupervised learning (UL) is also called *self-organizing learning* (without an external teacher or critic*)).

UL possible applications are [2]: pattern configuration, principal components analysis, classification of problems and vectors quantization, searching for a prototype, coding and decoding of signals, computation of feature maps.

1.3.1. Learning rules

1.3.1.1. Hebbian learning rule

The oldest and most famous of all learning rules is based on Hebb's postulate of learning [2]. The rule named in honour of the neuropsychologist Hebb (1949) can be expressed in the form:

$$\Delta w_{ij}(s) = \eta \, y_i(s) \, x_j(s) \,, \qquad (1.54)$$

where: η — learning rate.

The Hebbian rule is impractical for computations since it leads to an exponential growth of the synaptic weights w_{ij}. To avoid such a situation a forgetting factor is to be introduced to formula (1.54), (Kohonen, 1988):

$$\Delta w_{ij}(s) = \eta \, y_i(s) \, x_j(s) - \alpha \, y_i(s) \, w_{ij}(s) \,, \qquad (1.55)$$

where: α — new positive constant.

1.3.1.2. Oja's and Sanger's rules

Stabilization of the Hebbian rule by the forgetting factor cannot be achieved in case of linear activation functions. One of the possible rules which enables us to avoid the instability of weights computations is the rule by Oja (1982), cf. [19]:

$$\Delta w_{ij} = \eta \, y_i \left(x_j - y_i w_{ij} \right) \,. \qquad (1.56)$$

---

*) The term „*critic*" is used in *reinforcement learning* which can be treated as a kind of supervised learning. In the reinforcement learning „a critic" evaluates the tendency of the network to change one state to the other [2].

A modification of rule (56) was given by Sanger (1989), cf. [17,19]:

$$\Delta w_{ij} = \eta \, y_i \left(x_j - \sum_{k=1}^{i} y_k w_{ij} \right) =$$

$$= \eta \, y_i \left[\left(x_j - \sum_{k=1}^{i-1} y_k w_{kj} \right) - y_i w_{ij} \right] = \eta \, y_i \left(x'_j - y_i w_{ij} \right) \qquad (1.57)$$

Rule (1.57) is nonlocal since weight w_{ij} depends on the outputs and weights of other neurons.

In Fig.1.23 the example corresponds to 1000 two dimensional patterns (ξ_1, ξ_2) randomly selected from the two dimensional Gauss distribution $P(\xi)$. In Fig.1.23a the position of vector w is shown after $s = 2500$ iterations and in Fig.1.23b after $s = 1000$. Thin lines correspond to the learning trajectories at $\eta = 0.1$.

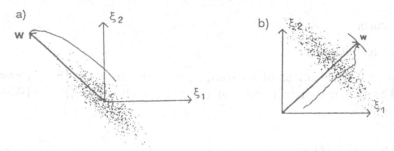

Fig.1.23: Unsupervised learning according to Oja's rule

It was proved, cf. [17], that the weight vectors tend to be of unit length and they are orthogonal to each other.

1.3.1.3. Grossberg's Instar and Outstar

Disadvantages of Hebb's type rules induced Grossberg (1982) to formulate his models of *stars*: *Instar* for input and *Outstar* for output, respectively. Instar (I) is the centre of coincidence of feedforward connections, and Outstar (O) is the centre of signal distribution, generated by the neuron, cf.Fig.1.24.

Instar

The Instar model reflects two types of activities: i) due to excitation by input signals, ii) relaxation to the basic state. This leads to the *Instar formula* [20,24]:

$$\Delta w_{ij}(s) = \eta \, y_i(s) \left[\bar{x}_j - w_{ij}(s) \right] , \qquad (1.58)$$

where: $\bar{x}_j = x_j / \sqrt{\sum_j x_j^2}$ – normalized components of the input vector.

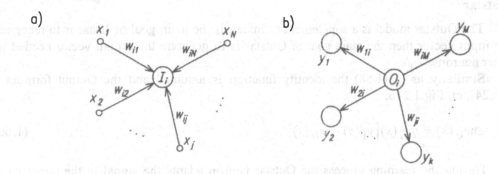

Fig. 1.24: Models of Instar (I) and Oustar (O)

Let us consider a special case of the identity activation function, i.e. $\beta = 1$ in function (1.7a), and zero value of the bias, i.e. $b_i = 0$. In this case the Instar formula can be written in the following vector form, corresponding to the Instar I_i.

$$w_i(s+1) = w_i(s) + \eta u_i^{(p)}(s)[\bar{x}^{(p)} - w_i(s)], \quad \text{where:} \quad u_i^{(p)}(s) = w_i^T(s)\bar{x}^{(p)}. \quad (1.59)$$

A simple interpretation of formula (1.59) is shown in Fig.1.25a for a single pattern $\bar{x}^{(p)} = \bar{x}$. In case of $u_i \to 1$ it is clear that $w_i(s) \to x^{(p)}$ for an increasing number of the iteration steps s.

The Instar neuron can be used in a simple network shown in Fig.1.25b for the *vector classification*. The weight vector $w^{(p)}$ is a representative of the class C_i for $i = 1,...,M$. If a vector $\bar{x}^{(t)}$, different than vectors $\bar{x}^{(p)}$ used for training, is presented to the network ready for operation then it can be qualified to a class C_k on the basis of the highest value of the scalar product $u_k^{(t)} = w_k^T\bar{x}^{(t)}$.

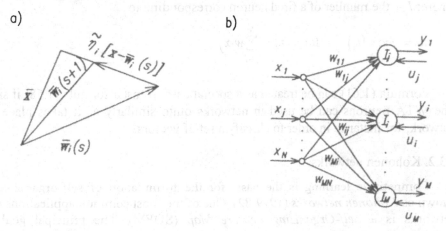

Fig. 1.25: a) Vector interpretation of formula (1.59), b) Instar one-layered network

Outstar

The Outstar model is a supplement to Instar. If the main goal of Instar is to recognize the input vector then the main goal of Outstar is to generate the output vector needed for other neurons.

Similarly as in (1.58) the identity function is assumed and the Output formula is [20,24], cf. Fig.1.24b:

$$\Delta w_{ki}(s) = \eta\, u_i(s)\big[y_k(s) - w_{ki}(s)\big]\,. \tag{1.60}$$

During the learning process the Outstar neuron adapts the signal to the target value $y_k(s) \to t_k(s)$.

The significant differences between the Instar and Outstar learning and BP learning discussed in Section 2 are worth emphasizing. In the error backpropagation the pattern was composed of a pair of vectors $(x,\ t)$. In cases of I or O neurons they are adapted either to vectors x or t under control of unsupervised learning.

1.3.1.4. WTA model

The term WTA stands for a '*winner-takes-all*' neuron. The WTA neuron is a neuron which wins the competition with other neurons for being the one to be active (fired). Thus, while in a network based on Hebbian-type learning several output neurons may be active simultaneously, in case of *competitive learning* only a single output neuron, i.e. WTA neuron, is active at any time.

According to Kohonen (1988) the *standard competitive learning rule* is defined by:

$$\Delta w_{ij}(s) = \begin{cases} \eta\big[\overline{x}_j - w_{ij}(s)\big] & \text{for} \quad i = I\,, \\ 0 & \text{for} \quad i \neq I\,, \end{cases} \tag{1.61}$$

where: $I -$ the number of a fired neuron corresponding to

$$u_I = \max_{i=1,\dots,M}(u_i) \qquad \text{for} \qquad u_i = \sum_{j=1}^{N} w_j x_j\,. \tag{1.61a}$$

Formula (1.61) can be treated as a special case of Instar formula (1.58) if signal $y_I = 1$. The WTA neurons can be used in networks quite similarly as it takes place with Instar network, for instance in order to classify a set of vectors.

1.3.2. Kohonen networks

Competitive learning is the basis for the formulation of self-organizing networks, known as *Kohonen networks* (1979-82). One of the most common applications of Kohonen networks is a *Self-Organizing Feature Map* (SOFM). The principal goal of SOFM algorithm is to transform an incoming signal pattern of arbitrary dimension into a one- or two-dimensional discrete map, and perform this transformation adaptively in a topological ordered fashion [2].

In. Fig.1.26 a scheme of transformation of a 2D vector $x = \{x_1, x_2\}$ into point i on 2D Kohonen map is shown.

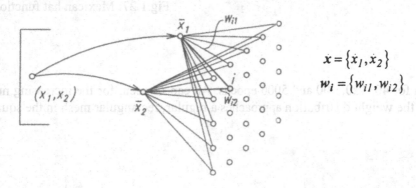

$$x = \{\dot{x}_1, \dot{x}_2\}$$
$$w_i = \{w_{i1}, w_{i2}\}$$

Fig.1.26: Transformation of vector x into a self-organizing feature map vector w_i

In Kohonen network a *modified competitive learning rule* is used:

$$\Delta w_{ij}(s) = \begin{cases} \eta(s)\zeta(i,I)[x_j - w_{ij}(s)] & \text{for} \quad i \in \Omega_I, \\ 0 & \text{for} \quad i \notin \Omega_I, \end{cases} \qquad (1.62)$$

where: Ω_I – neighbourhood of the winner I.

The modified formula (1.62) is assosiated with the activation of not only winner neuron I but also with neurons fired in winner neighbourhood Ω_I. The learning is conducted according to interconnection function $\zeta(i,I)$. Below function $\zeta(i,I)$ is defined as the so-called *Mexican hat function:*

$$\zeta(i,I) = \begin{cases} 1 & \text{for} \quad r_{iI} = 0, \\ \sin(ar_{iI})/(ar_{iI}) & \text{for} \quad |r_{iI}| \in (0, 2\pi/a), \\ 0 & \text{otherwise}, \end{cases} \qquad (1.63)$$

where: a – distance parameter.

In Fig.1.27 the graphics of the Mexican hat function id depicted as $\zeta(d)$, where: $d = ar_{iI}/2\pi$ According to this Figure region ① corresponds to a short-range lateral excitation, and region ② is associated with a penumbra of inhibitory action [2].

An application of Kohonen network is shown in Fig.1.28. This is an example of Demo SM2 taken from [20]. 1000 points (x_1, x_2) were randomly selected from a uniformly distributed pattern space, cf.Fig.1.28a. The feature mapping was made for a rectangular SOFM array. The network was composed of 6x5 = 30 neurons and the weight distribution

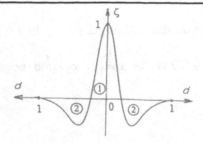

Fig.1.27: Mexican hat function

is shown for 40, 120, 500 and 5000 epochs. As can be seen, for the increasing numbers of iteration the weight distribution approaches a regular, rectangular mesh in the square.

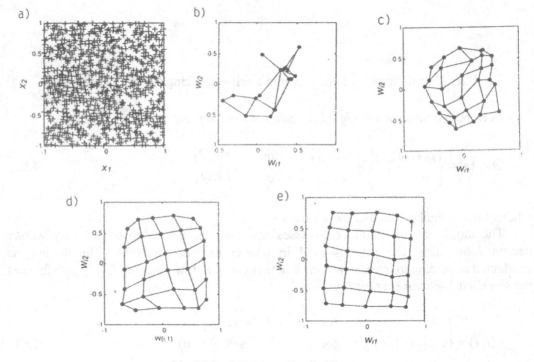

Fig.1.28: a) Pattern distribution,
b,c,d,e) Feature maps after 40, 120, 500 and 5000 epochs, respectively

The self-organizing feature map (SOFM) reflects the density of presented patterns $x^{(p)}$. SOFMs depend also on the shape of an assumed mask of maps (definition of neuron neighbourhood). The idea of SOFMs has been inspired by research on the human brain, where the cortical maps are sketched during the early development of the nervous system [2].

Kohonen networks are widely applied in mathematics for vector quantization [1]. Among engineering applications the image processing, damage detection of systems and fault diagnosis are worth mentioning [2].

Kohonen networks were used in [26] for the finite element mesh generation, cf.[10].

1.3.3. ART1 network

If we want to learn the neural network by new patterns these patterns can change or even „wash away" the information stored previously. This is associated with loss of NN *stability*. On the other hand, we would like the network to be adaptable in the sense of *plasticity*, i.e. the network should be plastic enough to be able to recognise and learn new information. These two conflicts are what Stephen Grossberg calls *stability-plasticity dilemma:* how can a network retain its stability while still being plastic enough to gainfully adapt in a changing environment [1].

The *Adaptive Resonance Theory* (ART) and ART networks were formulated to overcome Grossberg's dilemma. During the mid-1970s through 80s ART net architectures were developed by Grossberg and his associates. The simplest class of ART networks is ART1 implemented for binary inputs only. The ART2 network was designed for continuous inputs (Carpenter and Grossberg, 1987). In what follows only the ART1 network is discussed.

ART1 is composed of four components, Fig.1.29:

1) input/comparison layer F1,
2) output recognition layer F2,
3) attention subsystem A,
4) control subsystems G1 and G2.

The main components are layers F1 and F2. The layers are joined by a forward connection with weight matrix $W = \{w_j\}$ and the backward connections of the matix $V = \{v_{ij}\}$, where: $i = 1,...,N$, $j = 1,...,M$ are the numbers of neurons in layers F1 and F2, respectively.

Let us introduce an input binary vector x. At the output of layer F1 a copy of y is computed which is input to layer F2. In the competitive layer F2 only one, WAT neuron is activated and this pattern z is transmitted back from F2 in order to activate the layer F1. When vectors x and y are harmonized (nearly all bites of copy y equal the bites of the input vector x) the *Adaptive Resonance* (AR) takes place. That means that pattern x was recognized and classified by the winner neuron of layer F2.

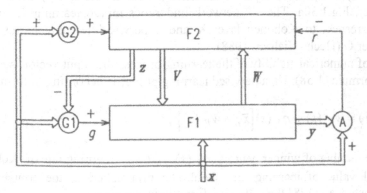

Fig.1.29: Components of ART1 network

When AR is activated vectors x and y are close to each other and attention subsystem A blocks layer F2 by means of reset parameter $r = 0$. The weights of the interlayer connections are then updated to have higher sensitivity of the network for the presented pattern.

When the adaptive resonance is nonactive the subsystem A implies $r = 1$ and output vector of layer F2 is cancelled, i.e. $z = 0$. In such a situation the control gain G1 sends signal $g = 1$ which makes $y = x$ and layer F2 is activated. The unsupervised learning starts and it is stopped after the input pattern is mapped to a class associated with layer F2.

The third case corresponds to the absence of a class to which the input vector could be mapped. The subsystems A, G1 and G2 enable us to activate a new winner of layer F1 to create a new class.

The first ART architectures were very unstable for the learning process. That is why control gains G1 and G2 were specified. Gain G2 simulates a theshhold to activate layer F2. The role of gain G2 is more complex. It works according to the 2/3 rule and affects the output of layer F1 if two from among three inputs x, G1 and F2 are active.

The above discussion is only an outline of ART algotihm. A precise description of the algorithm is given in [2,27].

ART1 network is of a complex architecture. In order to recognize a class of patterns there are needed $2N$ connections. The number of connections grows significantly since the attention and control subsystems have to be designed.

ART1 network is applied for character recognition, fault diagnosis, data fusion and tracking, process monitoring and control [2].

The ART network was also used as an associative memory device in the conceptual design of structural systems, cf.[8].

The main ART disadvantage is the network sensitivity to pattern imperfections. For instance, in the character recognition there are small disturbances which can lead to their false classification or creation of new uncommon classes.

1.3.4. CP network

The *CounterPropagation Neural Network* (CPNN) is a hybrid type network composed of two layers: 1) competitive layer of Kohonen's neurons K_r and 2) Grossberg layer of neurons G_i, cf. Fig.1.30a. The CP network undergoes mixed learning, i.e. unsupervised learning with respect to Kohonen layer K and supervised learning corresponds to the Grossberg layer G (Hecht-Nielsen, 1988).

Because of numerical stability of the learning process the input vector is normalized as in the Instar formula (1.58). Unsupervised learning proceeds according to formula (1.61):

$$w_{Rj}(s+1) = w_{Rj}(s) + \eta_K(s)\left[\bar{x}_j - w_{Rj}(s)\right], \tag{1.64}$$

where: R – the number of winner neuron, $\eta_K(s) \in (0,1)$ – learning rate of Kohonen layer.

An initial value of learning rate should be diminished in the course of iteration process. For instance in [28] the following formula is suggested:

$$\eta_K(s) = \frac{1}{(s+1)^2} , \qquad (1.65)$$

and the initial weights are assumed:

$$w_{rj}(1) = 1/\sqrt{N} . \qquad (1.66)$$

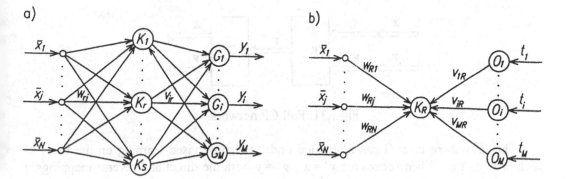

a)   b)

Fig.1.30: a) CounterPropagation neural network,
b) Learning of CPNN with Outstar output neurons

The Grossberg layer weights can be learned according to delta rule (1.16), modified by the unit signal from winner neuron K_R:

$$v_{iR}(s+1) = v_{iR}(s) + \eta_G(s)[t_i - y_i(s)] . \qquad (1.67)$$

In case of the identity activation function formula (1.67) takes the form of the Outstar formula (1.60), cf. Fig.1.30b:

$$v_{iR}(s+1) = v_{iR}(s) + \eta_G(s)[t_i - v_{iR}(s)] , \qquad (1.67a)$$

where: $\eta_G(s) \in (0,1)$ – learning rate of the Grossberg layer. In [28] formula (1.65) is also used for updating of $\eta_G(s)$.

The Kohonen layer K offers a possibility of separation of the input patterns on S classes, where: S – number of neurons in layer K. This implies the number of training patterns $L \leq S$. In case of nonhomogeneous patterns the number of neurons K_r can approach the number of patterns L.

This disadvantage of CP network was discussed in papers [29,30], where compression of the number of neurons in Kohonen layer was proposed.

CPNN can be used for mapping $f: \mathcal{R}^N \rightarrow \mathcal{R}^M$ but it operates as a look-up table. This means that for an input vector x from r-th class the output $y^{(p)}$ can be obtained. For a continuous mapping f an additional interpolation type postprocessing should be performed, cf. [3,28].

The CP network shown in Fig.1.30 can be called *Forward-Only CP Network* [3]. A generalization of this network is *Full CP Network*, cf. Fig.1.31. In case of Full CPNN both the input and output vectors x and y are presented to Kohonen layer K. After unsupervised learning the patterns can be separated up to S classes, where: S – number of neurons in layer K.

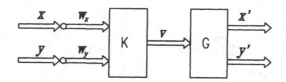

Fig.1.31: Full CP network

The Grossberg layer G can be trained under the supervised learning on the basis of vectors x' and y'. When vectors are $x' = x$, $y' = y$ both the direct and reverse mappings, f and f^{-1} respectively, are performed. An interesting operation of such a network can be made. If after the learning process the presentation of vector x or only y is performed then it induces both outputs x' and y'.

CPNNs can be trained more rapidly in comparison with BP networks, cf. [28], but generalization properties of CP networks can be limited significantly. That is why modifications proposed in [28,29,30] are worth introducing.

1.4. RECURRENT NETWORKS

1.4.1. Neural networks and associative memories

An *associative memory* corresponds to the recognition of previously learned patterns, even when some noise has been added. Usually patterns are composed of input and output vectors (x^p, t^p) or only of input vectors x^p, where: $p = 1,...,P$ and P is number of patterns.

We can distinguish three types of *associative networks* [2], cf.Fig.1.32:

1) *heteroassociative network* maps input vector $x^p \in \mathcal{R}^N$ to output target vector $t^p \in \mathcal{R}^M$;

2) *autoassociative network* is a special type of heteroassociative networks in which each output vector is associated with input vector, i.e. $t^p = x^p \in \mathcal{R}^N$;

3) *pattern recognition network* associates input vector x^p with scalar p.

From the viewpoint of the above classification the feedforward NNs are heteroassociative networks. The networks discussed in the frame of unsupervised learning are used in general for classification problems and these networks correspond to pattern recognition networks.

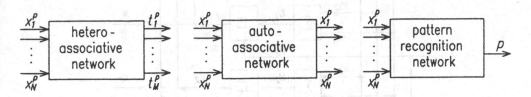

Fig.1.32: Types of associative networks

In case of a great number of patterns we can select a smaller number of *representative patterns* $A < P$ which can be related to classes or clusters. It is similar in case of splitting a set of patterns into learning and training sets. The learning set can be treated as a representative and generalization feature of feedforward networks can be measured as a „distance" between representative patterns and testing ones, i.e. noisy patterns.

Full CP network is an example of an autoassociative memory network in case $x = x'$, $y = y'$, cf. Fig.1.31. The autoassociative feature of this CP network is forced by such a mixed type of learning and presentation of the same patterns to the input and output neurons.

Autoassociative networks can also be designed by means of a special kind of *feedback connections*, i.e. the output of network is used as a new input, Fig.1.33. This corresponds to *reccurent networks* since in the iteration process there is $x(s) \rightarrow y(s) = x(s + 1)$.

The recurrent network can be called a *dynamic network* since in fact it behaves as a first-order dynamic system, i.e. each new state $x(t + 1)$ is completely determined by its most recent predecessor $x(t)$. Feedforward networks have no feedbacks so they can be classified as *static networks*.

1.4.2. Discrete Hopfield network

Let us consider the network shown in Fig.1.33. In the network there is no connection of a neuron with itself and the symmetry of the other weights is assumed:

$$w_{ii} = 0, \quad w_{ij} = w_{ij} \quad \text{for } i \neq j. \tag{1.68}$$

The scheme in Fig.1.33 takes into account the assumptions shown in (1.68). In the network the layered structure disappears in fact and the term „layer" can be used only in a conventional sense.

There are two phases of the operation of Hopfield neural network (HNN), namely:

1) in the *storage phase* a set of vectors, i.e. N-dimensional binary words, is memorised. These P vectors are called *fundamental memories*. In the storage phase the synaptic weights w_{ij} are computed applying a learning rule;

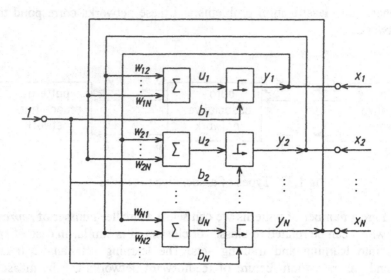

Fig.1.33: Hopfield discrete network

2) during the *retrieval phase* vector x, called a *probe* is imposed on HNN. Typically the probe represents an incomplete or noisy version of the fundamental memory of the network. In the retrieval phase vector x is associated with one of the fundamental memories x^P which was stored in the network.

In the storage phase the Hebbian learning method is usually used. Originally, in 1982, Hopfield considered the binary input variables which imply binary activation, as shown in Fig.1.33. Thresholds $\theta_i = -b_i$ are fixed externally to neurons $i = 1,...,N$ and for binary representation the synaptic weights are computed according to the following formula [3]:

$$w_{ij} = \begin{cases} \sum_{p=1}^{P}(2x_i^P - 1)(2x_j^P - 1) & \text{for} \quad i \neq j \,, \\ 0 & \text{for} \quad i = j \,, \end{cases} \qquad (1.69)$$

In case of bipolar representation (Hopfield, 1984) the learning formula is

$$w_{ii} = 0, \quad w_{ij} = \alpha \sum_{p=1}^{P} x_i^P x_j^P \quad \text{for} \quad i \neq j \,. \qquad (1.70)$$

Parameter $\alpha > 0$ can be assumed, e.g. $\alpha = 1/N$ in order to simplify the mathematical description of information retrieval, cf. [2]. In what follows the binary digits are used.

In the retrieval phase a probe is presented to the network and the activation potential of the i-th neuron $v_i(s)$ and output $y_i(s)$ is computed:

$$v_i(s) = \sum_{j=1}^{N} w_{ij} y_j(s) + b_i = u_i - \theta_i \quad \rightarrow \quad y_i(s) = F(v_i(s)). \tag{1.71}$$

The updating procedure is performed according to the formula:

$$y_i(s+1) = \begin{cases} 1 & \text{for} \quad u_i > \theta_i, \\ y_i(s) & \text{for} \quad u_i = \theta_i, \\ 0 & \text{for} \quad u_i < \theta_i. \end{cases} \tag{1.72}$$

The *asynchronous* updating is preferred, i.e. the iteration procedure is performed for a randomly selected neuron i. The updating is continued until there are no further changes to report, i.e.:

$$y_i(s+1) = y_i(s) \quad \text{for all} \quad i. \tag{1.73}$$

Fulfilling condition (1.73) corresponds to the *stable state* of the network. With the network stable state *atractors* a^r are associated. They are vectors of the input space which attracts the probe vectors presented to the trained network. In Fig.1.34 the attraction of vector x' to a^2 is shown

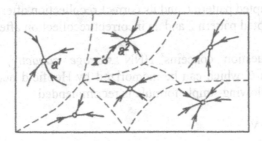

Fig.1.34: Domains (basins) of attraction

When the patterns are orthogonal to each other, i.e. $(x^k)^T x^l = 0$ for $k \neq l$, the attractors coincide with the fundamental memories, i.e. with vectors x^P from the storage phase as well as with their complements $\bar{x}^P$. Usually patterns x^P are correlated, so also *spurious attractors* are created. Despite this disadvantage HNNs are frequently used for recognition of corrupted or incomplete patterns.

Fig.1.35 corresponds to a computer experiment discussed in [2]. Hopfield network used in the experiment consisted of $N = 120$ neurons. Eight patterns were used as fundamental memories in the storage (learning) phase. From among those memories patterns 2 and 6 are depicted in Fig.1.35a. Each pattern contains 120 components (pixels) of the input vector, where black pixels correspond to 1 and white pixels to 0.

Figures 1.35b,c show correct and incorect recollections of corrupted patterns 6 and 2, respectively.

The example reflects the error-correction capability of the network. The corruption in Fig.1.35b seems to be more serious (36 wrong pixels) than that in Fig.1.35c (26 incorrect pixels) but both probes are attracted to the same fundamental memory 6.

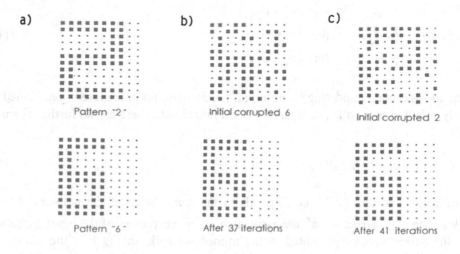

Fig.1.35: a) Two patterns from fundamental memories,
b) Initial corrupted pattern 6 and its correct recollection after 47 iterations,
c) Initial corrupted pattern 2 and an incorrect recollection after 41 iterations

An important question concerns HNN *storage capacity*, i.e. the number of fundamental memories C which can be memorized by Hopfield network composed of N neurons. In [19] the following simple formula is recommended

$$C = P_{max} \approx 0.138\,N \ . \tag{1.74}$$

The storage capacity estimation was extensively discussed in literature, cf. [2,17].

1.4.3. Energy function and continuous Hopfield networks

For Hopfield network the *energy function* is defined:

$$E(y(s)) = -\frac{1}{2} y^T(s)\, W\, y(s) + \theta^T y(s) = -\frac{1}{2} \sum_{i=1}^{N} \sum_{j=1}^{N} w_{ij} y_i(s) y_j(s) + \sum_{i=1}^{N} \theta_i y_i(s). \tag{1.75}$$

Function E corresponds to *Lyapunov function* which is used in the stability analysis of dynamic systems. The energy changes during the information retrieval and tends to a minimal value which corresponds to a stable equilibrium state of the network:

$$E(s+1) = E(s) \quad \Leftrightarrow \quad y(s+1) = y(s). \tag{1.76}$$

In Fig.1.36 a mechanical analogy of the retrievial process is shown. Starting from an initial state $x = y(s)$ the iteration process tends to a local minimum of energy function E which corresponds to an attractor a^r. A region associated with this attractor is called the *basin of attraction*.

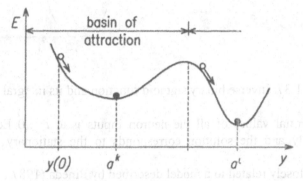

Fig.1.36: Graphic interpretation of dynamic processes in Hopfield network

Besides discrete HNN also *continuous* (analog) version of *Hopfield network* was formulated (Hopfield, Tank, 1985). In case of the continuous Hopfield network (CHN) variables y_i are real numbers and energy function is defined in the following form:

$$E(y) = -\frac{1}{2}\sum_{i=1}^{N}\sum_{j=1}^{N}w_{ij}y_iy_j + \sum_i\theta_iy_i + \sum_i\int_0^{y_i}F_i^{-1}(\eta)d\eta , \tag{1.77}$$

where: F_i – activation function of the i-th neuron.

In comparison with energy function (1.75) defined for discrete HNN an additional term is added in (1.77) for continuous Hopfield network. This term is associated with the assumed activation function.

The added term acts as a bareer function, i.e. it constrains values of y_i to the range $(0,1)$. In Fig.1.37 the inverse binary sigmoid function $F_\sigma^{-1}(y)$ is presented for $F_\sigma = 1/[1+\exp(-\sigma y)]$ where $\sigma > 0$. The integral of this activation function $I(y)$, computed within the interval $[0.5, y)$ is also shown in Fig. 1.37.

The dynamics of the analog version of HNN is described by a set of ordinary equations:

$$\tau_i \frac{dv_i}{dt} = -\frac{\partial E}{\partial y_i}\bigg|_{y_i(t)=F_i(v_i(t))} \equiv -v_i + \sum_{j=1}^{N} w_{ij} y_j - \theta_i \quad \text{for} \quad i = 1,...,N, \tag{1.78}$$

where: $\tau_i > 0$ – time constant, t – time type variable.

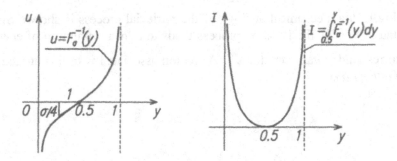

Fig.1.37: Inverse binary sigmoid function and its integral

Assuming the initial values of all the neuron inputs v_i at $t = 0$ Eqs (1.78) can be integrated numerically and the solution corresponds to the stationary value of energy (1.76).

Eqs. (1.78) are closely related to a model described by Pineda (1987), cf. [2]:

$$\tau_i \frac{dy_i}{dt} = -y_j(t) + f\left(\sum_i w_{ji} y_i\right) - \kappa_j \quad \text{for} \quad j = 1,...,N. \tag{1.79}$$

Multiplying both sides of Eq. (1.79) by w_{ij}, summing with respect to j and then substituting the transformation

$$u_i(t) = \sum_j w_{ij} y_j(t) \tag{1.80}$$

the following equation is obtained

$$\tau_i \frac{du_i}{dt} = -u_i(t) + \sum_j w_{ij} f(u_i) + q_i , \tag{1.81}$$

where the biases in (1.81) are

$$q_i = -\sum_j w_{ij} \kappa_j . \tag{1.82}$$

Continuous HNNs are used as a means of solving optimization problems. It is exactly in this field Hopfield networks are applied to the anaysis of various problems of optimum design in mechanics, cf. [10]. These problems are also discussed in Chapter 6.

1.4.4. Circuit model of Hopfield continuous network

The analog neural network can be implemented by means of an electronic elements network shown in Fig. 1.38. The network has N amplifiers (e.g. nonlinear amplifiers of sigmoid characteristics can be used) and the elements of the following characteristics: r_i, R_{ij} – input and synapses resistancies, C_i – capacity, I_i – externally supplied input current, v_i, V_i – input and output voltages of the i-th amplifier.

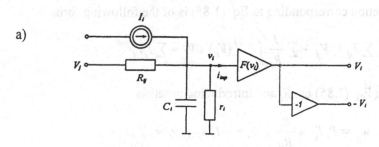

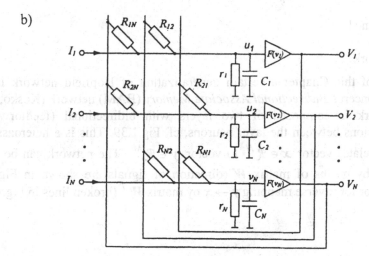

Fig.1.38: Schemes of circuit model of Hopfield network:
a) Single amplifier, b) Resistor-capacitance model of the network

Applying Kirchhoff's current law with respect to the input node of the network in Fig.1.38 we get

$$C_i \frac{dv_i}{dt} = -\frac{1}{r_i}v_i + (V_j - v_j) + I_i = -\left(\frac{1}{r_i} + \sum_j \frac{1}{R_{ij}}\right)v_i + \sum_j \frac{1}{R_{ij}}F(v_j) + I_i \ . \qquad (1.83)$$

After introducing the notation

$$\frac{1}{R_i} = \frac{1}{r_i} + \sum_j \frac{1}{R_{ij}} , \qquad T_{ij} = \frac{1}{R_{ij}} ,$$ (1.84)

the circuit equation takes the form:

$$C_i \frac{dv_i}{dt} = -\frac{v_i}{R_i} + \sum_j T_{ij} V_j(t) + I_i ,$$ (1.85)

$$V_j(t) = F_j\big(v_j(t)\big) \qquad \text{for} \quad i,j = 1,...,N .$$

Energy function corresponding to Eq. (1.85) is of the following form

$$E = -\frac{1}{2} \sum_i \sum_j T_{ij} V_i V_j + \sum_i \frac{1}{R_i} \int_{V_0}^{V} F^{-1}(V_i)\, dV_i - \sum_i I_i V_i .$$ (1.86)

Multiplying Eq. (1.85) by R_i and introducing notation

$$\tau_i = C_i R_i , \quad w_{ij} = R_i T_{ij} = \frac{R_i}{R_{ij}} , \quad \theta_i = -I_i R_i , \quad y_i = V_i ,$$ (1.87)

Eq. (1.87) is obtained.

1.4.5. BAM network

At the end of this Chapter a certain generalization of Hopfield network is briefly discussed. This concerns *Bidirectional Associate Memory* (BAM) network (Kosko, 1987).

BAM network is composed of two layers with bidirectional (feedforward and feedback) connections between the layer neurons, cf. Fig.1.39. This is a heteroassociative network since it relates vector $x \in \mathcal{R}^N$ to vector $y \in \mathcal{R}^M$. The network can be used for mapping $x \to y$ by means of matrix W (direction of signals are shown in Fig.1.39 as continuous lines) or for inverse mapping $y \to x$ by matrix W^T (broken lines in Fig. 1.39).

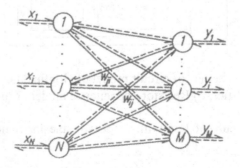

Fig.1.39: A scheme of BAM network

Let us assume that discrete BAM network has bipolar activation functions. In the storage phase the patterns of fundamental memory are used for the computation of weights, according to Hebbian rule:

$$w_{ij} = \sum_{p=1}^{P} y_i^P \, x_j^P \, .$$ (1.88)

Matrix $w = [w_{ij}]$ is a long term memory and it can reproduce the patterns according to relations:

$$y^P = sgn(W \, x^P), \quad x^P = sgn(W^T y^P) \, .$$ (1.89)

where sgn corresponds to the action of bipolar step activation function with respect to each component of vectors y^P, x^P, respectively.

In the retrievial phase probe x' is converged to memory y^r and vice versa:

$$sgn(W \, x') \rightarrow y^r, \quad sgn(W^T y') \rightarrow x^r .$$ (1.90)

When the weight matrix is quadratic and symmetric, i.e. $W = W^T$, and the input vectors are the same, i.e. $x = y$, BAM network is reduced to Hopfield network (HNN).

Similarly as in case of HNN, BAM network has the storage capacity [17]:

$$C = \frac{n}{2 \log_2 n},$$ (1.91)

where $n = \min(N, M)$.

BAM network can be designed in discrete, binary and bipolar versions, as well as continuous BAM [3]. From among many applications also the use of a BAM network to the structural analysis by the stiffness method can be quoted [32].

ACKNOWLEDGEMENT

Financial support by the Polish Committee for Scientific Research, Grant No 8 T11F 022 14 „Artificial neural networks in structural mechanics and bone biomechanics", is gratefully acknowledged.

REFERENCES

1. Rojas, R.: Neural Networks – A Systematic Introduction, Springer, Berlin/ Heidelberg 1966.

2. Haykin, S.: Neural Networks – A Comprehensive Foundations, Macmillan College Publishing Co., Englewood Cliffs, NJ 1994.

3. Fausett, L.: Fundamentals of Neural Networks – Architectures, Algorithms, and Applications, Prentice Hall Inc., Englewood Cliffs, NJ 1994.

4. Rumelhard, D.E. and McClelland, J.L.(Eds): Parallel Distributed Processing – Explorations in the Microstructure of Cognition, Vol.1, MIT Press, Cambridge, MA 1986.

5. Adeli, H. and Yeh, C.: Perceptron learning in engineering design, Microcomputers in Civil Eng., 4 (1989), 247-256.

6. Topping, B.H.V. and Khan, A.I.(Eds): Neural Networks and Combinatorial Optimization in Civil and Structural Engineering, Civil-Comp Press, Edinburgh 1993.

7. Panagiotopoulos, P.D. (Sci. Coordinator): Applications of the Neural Networks in Engineering and Medicine, STRIDE-Programm of the EEC, 1993.

8. Hajela, P. and Berke, L.: Neural networks in structural analysis and design – an overview, Computing Systems in Eng., 3(1992), 525-538.

9. Waszczyszyn, Z.: Some recent and current problems of neurocomputing in civil and structural engineering, in: Advances in Computational Structures Technology (Ed.B.H. V. Topping), Civil-Comp Press, Edinburgh 1996, 43-58.

10. Topping, B.H.V. and Bahreininejad, A.: Neural Computing for Structural Mechanics, Saxe-Coburg Publications, Edinburgh 1997.

11. Paez, Th. L.: Neural networks in mechanical system simulation, identification and assessment, Shock and Vibration, 1 (1993), 177-199.

12. Waszczyszyn, Z., Pabisek, E. and Mucha, G.: Hybrid neural network/ computational programs to the analysis of elastic-plastic structures, in: Discretizations Methods in Structural Mechanics (Eds. H.Mang and F.G. Rammerstorfer), Kluwer Acad. Publ., Dortrecht 1999, 189-198.

13. Gunaratnam, D.J. and Gero, J.S.: Effects of representation on the performance of neural networks in structural engineering applications, Microcomputers in Civil Eng., 9 (1994), 97-108.

14. Wu, X., Ghaboussi, J. and Garret, J.H.Jr.: Use of neural networks in detection of structural damage, Comp. Stru., 42 (1992), 649-659.

15. Stavroulakis, G.E. and Abdalla, K.M.: A systematic neural network classificator in mechanics – an application in semi-rigid steel joints, Eng. Analysis and Design, 1 (1994), 272-292.

16. Pao, V.H.: Adaptative Pattern Recognition and Neural Networks, Addison-Wesley, Reading MA 1996.

17. Osowski, S.: Neural Networks (in Polish), Oficyna Wyd. Polit. Warszawskiej, Warsaw 1994.

18. Hajela, P. and Berke, L.: Neurobiological computational models in structural analysis and design, Comp. Stru., 41 (1991), 657-667.

19. Hertz, J., Krogh, A. and Palmer, R.G.: Introduction to the Theory of Neural Computation, Addison-Wesley, Reading MA 1991.

20. Demuth, H. and Beale, M.: Neural Network Toolbox for Use with MATLAB – User's Guide, The Math Works, Natick MA 1994.

21. Zell, A. (Ed.): SNNS – Stuttgart Neural Network Simulator, User's Manual, Version 4.1, Univ. Stuttgart 1995.

22. Masters, T.: Practical Neural Network Recipes in C++, Academic Press 1993.

23. Waszczyszyn, Z., Pabisek, E. and Stankiewicz, P.: Application of back-propagation neural networks to the buckling analysis of cylindrical shells, in: Neural Networks and Their Applicatios, Proc. 3rd Conf., Częstochowa, Poland, 1997, 473-478.

24. Hegazy, T., Fazio, P. and Moselhi, O.: Developing practical neural network applications using back-propagation, Microcomputers in Civil Eng., 9 (1994), 145-159.

25. Cios, K.J., Vary, J., Berke, J. and Kautz, H.E.: Application of neural networks to prediction of advanced composite structures mechanical response and behaviour, Computing Systems in Eng., 3(1992), 539-544.

26. Sarzeaud, O., Stephan, Y. and Touzec, C.: Finite element meshing using Kohonen's self-organizing maps, in: Artificial Neural Networks (Eds. T.Kohonen et al), Elsevier Science Publishers, 1991, 1313-1317.

27. Braspenning, P.J., Thuijsman, F. and Weijters, A. J. M. M. (Eds): Artificial Neural Networks – An Introduction to ANN Theory and Practice, Springer, Berlin/ Heidelberg 1995.

28. Adeli, H. and Park, H.S.: Counterpropagation neural networks in structural engineering, J.of Structural Eng., 121 (1995), 1205-1212.

29. Szewczyk, Z.P. and Hajela, P.: Neural network approximation in a simulated annealing based optimal structural design, Structural Optimization, 5 (1993), 159-165.

30. Szewczyk, Z.P. and Hajela, P.: Damage detection in structures based on feature-sensitive neural networks, J. of Computing in Civil Eng., 8 (1993), 163-178.

31. Kortesis, S. and Panagiotopoulos, P.G.: Neural networks for computing in structural analysis – methods and prospects of applications, Int. J. Num. Methods in Eng., 36 (1993), 2305-2318.

32. Lou, K-N. and Perez, R.A.: On the artificial neural networks for structural analysis, in: Recent Advances in Structural Mechanics, ASME, PVP-295/NE-16, (1994), 149-161.

CHAPTER 2

GENETIC ALGORITHMS AND NEURAL NETWORKS

W.M. Jenkins
University of Leeds, Leeds, UK

ABSTRACT

The basic genetic algorithm is introduced including the representation of individuals in populations, data structures for the representation of variables, binary strings, assessment of individual fitness, selection for recombination, crossover and mutation operators. The penalty function method of handling design constraints is introduced. The basic GA is illustrated by optimizing a simple structural design. We consider how we might improve the GA by on-line adaptation of the main controls. We then review string coding, the schema theorem and the formation of building blocks in the strings. We consider the coding of continuous-valued variables and bit array representations, elitism, methods of maintaining diversity in the population and introduce a further illustration in structural optimization. The application of the genetic algorithm is then extended into large scale-situations, particularly design situations involving a large number of variables. A combinatorial space reduction heuristic based on a record of parameter selection intensities is described. The allocation of fitness to partial strings is reviewed. Consideration is given to the multi-objective GA and pareto optimality. There follows a brief introduction to mathematical models of the GA. The GA is used to train a neural network as an alternative to back-propagation. We consider the 'permutations' problem and introduce the concept of 'shift'. The method is illustrated by training a neural network for structural analysis. The chapter concludes with a brief review of the implicit parallelism of the GA and suggestions as to how the algorithm might be improved with parallel hardware and a further example application.

2.1. INTRODUCTION

The genetic algorithm (GA) is used in engineering problems requiring a powerful and robust computational device for searching large combinatorial spaces [1-7]. The algorithm processes a population of 'individuals', typically engineering designs, generation-by-generation operating on a set of coded values of the data. A 'fitness' measure is defined and applied to each individual, relatively fit individuals are combined to produce new individuals. Re-combination is designed to increase fitness so the maximum and average fitnesses in the population will rise, rapidly at first. The process continues until an ultimate maximum fitness is reached - this may or may not be the 'optimum'.

Briefly an 'optimum' design will be defined as a set of values of design variables that together minimize an 'objective' function. Usually the objective will be cost-based, however this is frequently difficult to achieve and other criteria, for example volume of material, are sometimes adopted for optimum seeking purposes. Practical considerations suggest that an insistence on reaching a global optimum is hardly justified in view of the uncertainties surrounding the cost and other data involved. The designer will generally be satisfied with a 'progression' towards an optimum accompanied by control facilities allowing parametric and comparative studies if desired.

Classical optimization methods involve primarily the exploitation of gradients and are deterministic in character. The GA has a stochastic nature since random components are introduced both in the initial generation and in the subsequent processing. The GA does not use gradient information in its search process, although that is not to say that useful gradients will not be indirectly exploited by the algorithm if they exist. The GA operates on 'discrete' values of data. This is appropriate since, for example, structural steel sections are discrete in properties, concrete slabs have discrete thicknesses since practical dimensions will be rounded and will vary at discrete intervals. Nothing is lost in generalization if we take all design variables to be discrete. If we meet genuinely continuous variables then these can usually be discretized without difficulty. A result of the assumption of discrete variables is that the design space formed by all combinations of the variables is finite and calculable in size. This space may be very large, a phenomenon termed 'combinatorial explosion' and this is a problem we shall consider in Section 20. We shall find that the GA requires some care in applying the various controls, is computationally demanding and requires considerable memory provision - these requirements are adequately provided by modern computing hardware.

It is necessary to provide means to handle infeasibility in designs caused by constraint violations. We shall use a 'penalty' method for this.

2.2. THE BASIC GA

We now undertake a brief review of the GA in its basic form.

2.2.1 Representation of variables – data structures

All data are processed in a general binary form. An individual (design) is represented by a binary string of appropriate length incorporating, generally by simple concatenation, values of the variables constituting the design. This string is called a 'chromosome' in the

GA. In the usual form of the GA, the bit string representing a single individual is handled as a complete entity and subjected to the GA operators, however, if we anticipate more complex design situations with large numbers of variables then it will be sensible to partition the individual string in some way and apply the GA processes to the sub-strings. This topic will be pursued in Point 2.2.6.

The total length of the bit string will depend on the number of design variables (n) and the number of discrete values to be used for each variable. Suppose we have for each variable a string of length L then the number of discrete values provided is,

$$v = 2^L \tag{2.1}$$

so if we use a string of length 5 bits, then we provide 32 discrete values for each variable and the total length of the string will be n*32. In what follows we shall assume that each variable may have a different length in the total string since clearly this will be a useful option to have.

The values of the variables appearing in the bit string(s) are coded so now we must touch on the means of achieving this coding. Suppose we have a string of 5 bits for a variable x_i then we provide 32 decoded values for this variable. These may be the available depths of a beam in the form of a one-dimensional array, or they may represent properties of steel sections (area, second moment of area, radius of gyration etc.,) in a two-dimensional array. Clearly the string length chosen must be large enough to contain all the values to be offered to the algorithm according to (2.1). If the requirement lies between one value of L and the next some values can be duplicated to fill-up spaces or some other device may be used.

2.2.2. Assessment of fitness

Purely for simplicity we may assume that the target in the optimization process is to find the structure of minimum volume of material. Now the action of the GA will be to increase fitness so we need to express the fitness of an individual in the form,

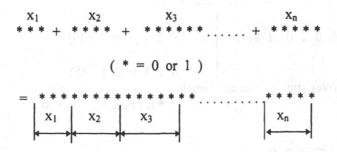

Fig. 2.1: Representation of a design by concatenation of binary strings

$$\text{fitness} = F = C - y \tag{2.2}$$

where y is the objective function (volume of material for convenience). It will be convenient to avoid the production of negative values of fitness and we can do this by setting the constant C as a large number, conveniently the maximum volume consistent with the data provided.

2..2.3. Composing the initial population

It is usual to determine a suitable population size and retain this throughout the processing. The population size should be large enough to provide the GA with adequate scope to search the combinatorial space, on the other hand the larger the population the more memory is required. In structures of moderate size, with strings lengths of the order of 6 or 7 for *each* variable, a population of 50 individuals has been found generally satisfactory. The initial population is generated by taking random coded values for each variable individual-by-individual until the population is filled. It is usual practice to have an even number of individuals in the population but this is not essential.

2.2.4. Selection for recombination

A number of methods exist for selecting individuals for re-combination - a process sometimes called 'breeding'. Whatever method is used, the intention is to give relatively fit individuals a high probability of being selected for re-combination. The method of 'stochastic remainder' is popular and reliable so we shall examine this method. For each individual the ratio, (fitness / average fitness) is formed. The selection weighting then attached to the individual is first the integer part of this ratio then increased probabilistically using the fractional part as probability. It follows that a highly fit individual may be selected more than once, a fit individual may be selected once and an unfit individual not at all.

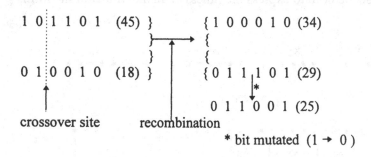

Fig.2.2: Crossover and mutation in binary strings

2.2.5. Recombination of individuals

The principal genetic operator used in the recombination of individuals is termed 'crossover'. A crossover site is chosen, usually randomly, in the two strings to be recombined and the bits to one side of the crossover site are exchanged between the two strings. The operation is illustrated in Fig 2. In general the decoded values of the two strings will be changed. Bearing in mind that both strings have been selected on the basis of fitness, then clearly there is a likelihood of one or both of the new strings possessing an even higher fitness. If not, then the new strings will simply go back into the pool and may or may not be selected for the next generation. Intuitively one can see that it is highly likely that progress in the direction of an increasing average fitness in the population will take place as better strings evolve, are captured and replace previous above average-strings. Crossover is carried out at a rate determined by a specified probability (Pc) usually of the order of 0.6 - 1.0.

The second genetic operator is termed 'mutation', this is applied by selecting one bit in a string, usually randomly, and flipping this bit (0 to 1, 1 to 0). Mutation is carried out at a specified probability, usually of the order of 0.001 - 0.01. Mutation can introduce fresh genetic information by encouraging diversity in the population whereas crossover cannot do this. Crossover is an 'exchange' operator and cannot produce new information ie if a 1 bit exists in a particular location in the ultimate design and if this 1 is not present in *any* of the individuals in the initial population, then it cannot be introduced by crossover, but could by mutation. Both operators are illustrated in Fig. 2.2. The probabilities controlling crossover and mutation can with advantage be adapted during the processing of the GA.

2.2.6. Alternative forms of crossover

Whatever form of crossover is used, a straight replacement of two parent strings by two child strings is frequently used. In Fig. 2.3(A), a simple one-point crossover in concatenated strings is shown. If these are long strings then most of the decoded values of the variables are unchanged by the process. An alternative, shown at (B), is to carry out crossover in each pair of substrings representing the variables. This is clearly convenient and much more disruptive, thus accelerating progress, however it requires that whatever fitness measure is applied to the complete string is also applied to the individual substrings and the justification for this is not immediately apparent. Some work has been done [25,34] which suggests that an overall fitness measure for the complete string can be applied to partial strings in this way. A further development can usefully be applied in the processing and that is to use randomly selected parents for each production of child strings representing the variables. This is shown at (C) in Fig. 2.3.

In 'uniform' crossover a, randomly created, crossover 'mask' is used, possibly applied to substrings. In this form of crossover if a 1 occurs in the mask the corresponding gene is taken from parent 1 whereas if a zero appears then the corresponding gene is taken from parent 2. A new random mask may then be created for the second child string. Some authors [12] confirmed that partitioning on the basis of substrings of individual variables produced the best results.

2.3. FITNESS SCALING

It is known that the operation of the GA can be improved by scaling the fitnesses, this has the effect of increasing the discriminating ability of the GA.   The method recommended by Goldberg [1] is normally used where relatively high fitnesses are increased and low fitnesses reduced whilst maintaining the same average for both raw and scaled fitnesses.

2.4. DESIGN CONSTRAINTS

Structural design is generally much affected by constraints; stresses, displacements, geometry and many other aspects of a structural design are subjected to constraints.   The treatment of design constraints is perhaps the most difficult area in structural optimization. In the penalty method, the fitness of a design is reduced when constraints are violated.

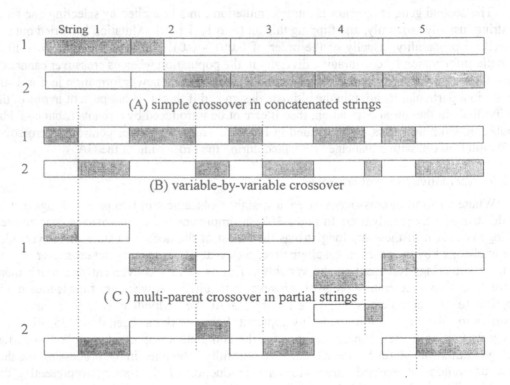

Fig. 2.3: Alternative forms of crossover. Child strings formed from parents 1 and 2 by crossover as shown.

When composing a penalty function it will be useful to build-in the idea of minor and major constraint violations to acknowledge the somewhat uncertain nature of the structural data; for example if a particular deflection must not exceed say 25 mm then we would impose a more modest penalty on the fitness if the actual displacement were 26 mm as compared with a displacement of 30 mm. We can build this idea into a penalty function,

$$\text{Penalty} = P = k\,(d)^m \tag{2.3}$$

where, $d = q/a$ for $q>a$ or a/q for $a>q$
and, a = actual value of the constrained quantity
             q = allowable value of the constrained quantity

We define,
     P_0 = penalty level negligible in comparison with C in (2.1)
     P_1 = lowest *significant* level of penalty
     d_0 = negligible penalty violation
     d_1 = lowest level of *significant* violation.

Substituting in (2.3) we obtain,

$$m = \text{Log}(P_0 /P_1)/ \text{Log}(d_0 /d_1) \tag{2.4}$$

and, $$k = P_0 /(d_0)^m \tag{2.5}$$

     As an example, suppose C = 100000 and we decide (this is a matter for the designer's judgement) to take d_0 = 1.01 (ie a 1% violation), d_1 = 1.2 (a 20% violation), P_0 = C/100 and P_1 = C/10 then we obtain, m = 13.4 and k = 875.   A suitable penalty function would then be,

$$P = 900 * (d)^{13} \tag{2.6}$$

It is desirable to relate the absolute values of the penalties to the maximum fitness ie C in (2.2). In the example above, a 44% violation (d=1.44) would apply a penalty approximately equal to C and effectively remove the individual.

2.5. A SIMPLE DESIGN OPTIMIZATION

     As an illustration of the application of the GA in simple form we take the problem of finding the optimum design of a simply-supported reinforced concrete beam. The problem has been examined by several authors including B. K. Chakrabarty using linear programming [8]. A GA solution was provided by Jenkins [9] as in the following outline.
     The beam is assumed rectangular in cross-section and singly reinforced. There are three design variables (cm units), x_1 = area of tension steel, x_2 = effective depth of beam, x_3 = width of beam. The data to be used in the design is,

     i) ultimate applied bending moment (excluding self-weight) = 687.5 kNm
     ii) design concrete strength = 30 N/mm^2 (f_c)
     iii) design steel stress = 300 N/mm^2 (f_y)
     iv) mass of concrete = 2323 kg/m^3
     v) unit costs of concrete, steel, shuttering = \$64.5 /m^3 , \$0.72 /kg, \$2.155 /m^2

The design equations used were,

total ultimate bending moment, $M = 687.5 + 0.03135 * x_2 * x_3$ (2.7)

$$M / (0.9 * f_y * x_1) + n/2 = x_2 \qquad\qquad (2.8)$$

from which n the neutral axis depth can be obtained.

The chosen objective function to be minimized is the cost per unit length of the beam (y$/cm),

$$y = 0.005652(x_1)+0.00007095(x_2)(x_3)+0.0004741(x_2)+0.0002155(x_3) \qquad (2.9)$$

The design constraints were,

$$x_1 <= 0.085 * x_3 * n \qquad\qquad (2.10)$$

$$\text{and} \quad x_3 / x_2 >= 0.4 \qquad\qquad (2.11)$$

(2.10) ensures that the beam is not over-reinforced and (2.11) applies a b/d restriction.

In running the GA, 31 (decoded) discrete values were provided for each of the three variables, as follows,

$x_1 = 30,31,32 \ldots\ldots 60$ (cm^2)
$x_2 = 60,61,62 \ldots\ldots 90$ (cm)
$x_3 = 20,21,22 \ldots\ldots 50$ (cm)

The procedure using the GA was as follows,

1) select values for the variables
2) compute M from (2.7)
3) compute n from (2.8)
4) compute fitness of each individual in the population
5) check design constraints and penalize as necessary.

If the restriction on the ratio b/d is withheld, the algorithm selects the maximum value of d and the minimum value of b as would be expected. The optimum design was, $x_1 = 35$ cm, $x_2 = 89$ cm and $x_3 = 20$ cm with the corresponding cost 0.371 $/cm. The GA required 10 generations of processing a population of 50 individuals to achieve this result.

When the b/d constraint was applied the optimum design was; $x_1 = 46$ cm, $x_2 = 70$ cm and $x_3 = 28$ cm with the corresponding cost 0.438 $/cm. The design space searched by the algorithm comprised 29792 ($=31^3$) combinations of the variables. The number of actual designs visited by the algorithm was 500 representing 1.68% of the total space.

We have applied the GA to a comparatively small problem in order to illustrate the principles. We go on to develop the algorithm for larger scale applications.

2.6. ENHANCING THE GA WITH CONTROL ADAPTIVITY

In setting values for the control probabilities for crossover (P_c) and mutation (P_m) it is generally accepted that P_c should be high and P_m low. It has been shown that an improvement in the operation of the algorithm is obtained if these controls are adapted during the process.   A simple but effective adaptation is obtained using the linear relationships,

$$Pc[i] = Pco + (1 - Pco) * r \qquad (2.12)$$

$$Pm[i] = Pmo * (1 - r) \qquad (2.13)$$

where,

   $Pc[i]$ = probability of crossover in the string representing variable i
   $Pm[i]$ = probability of mutation in the string representing variable i
   Pco, Pmo = initial values of control probabilities
   r = generation number / maximum generation number.

With these adaptive rules, Pc tends to unity and Pm tends to zero at the maximum generation number set by the user.   In most cases it will be found that the algorithm terminates before this situation is reached.

2.7. STRING CODING

The binary string coding we considered in Section 2.1 is commonly employed but has some drawbacks and there are alternatives. Parameter values which lie close in reality may be very distant in the binary string, for example decimal 32 is binary 100000 whereas decimal 31 is binary 011111 which is very different in the binary space. In the 'Gray' code adjacent integers all differ by one bit, a Hamming distance of 1, however string decoding is more difficult in the Gray system [3].

Other forms of coding are used when problems require sequence representation.

2.8. THE SCHEMA THEOREM – BUILDING BLOCKS

If we wish to analyse the fundamental behaviour of the GA to determine 'what makes it work' then a useful concept is that of the 'schema', it will be seen that   schemata are structured subsets of the combinatorial space, or subsets of 'similar' chromosomes (similarity templates). If we encounter the following strings,

        (a) 0 1 1 0 0 1 and (b)   1 1 0 0 1 1

then both (a) and (b) belong to a family of strings (schemata) described as,

            (c) * 1 * 0 * 1

in which the bits common to both (a) and (b) are specified in (c) whilst the unspecified bits are denoted * indicating that these bits can be either 1 or 0, the symbol * being treated as a 'wild card'. Schemata can be thought of as hyperplanes in n-dimensional space. Each chromosome is an instance of 2^n distinct schemata since each bit can take its actual value (0 or 1) or a wild card * . It follows that any given chromosome contains information relating to all schemata of which it is a member. In a population of P chromosomes the upper limit on the number of schemata represented is $P(2)^L$. This is an upper limit because there is likely to be some duplication. It follows that, as the GA processes the individual chromosomes, there is a parallel processing of schemata. This phenomenon is termed 'implicit parallelism'.

We now proceed to develop the 'schema theorem'. We define the terms 'schema length' and 'schema order' as follows,

schema length = d = number of bits from the first specified bit to the last specified bit,

schema order = o = number of specified bits.

We express first the probability of selection. The number of examples of schema S in population (t+1) will be a function of the number in population (t) and, the schema fitness F(S) and the average fitness of the population (F'),

$$n(S, t+1) = n(S, t) * F(S) / F' \tag{2.14}$$

We now seek to express the probability that a schema S will be destroyed by the operation of crossover. This is the product of the probabilities of crossover (Pc) and the probability that the crossover site lies within the defining length of the schema.

$$Pc\, d / (L - 1) \tag{2.15}$$

This must be multiplied by the probability that the other parent is also an instance of schema S

$$(1 - P[S, t]) \tag{2.16}$$

Hence the probability of survival of schema S is,

$$>= 1 - Pc\, d\ (1 - P[S,t]) / (L - 1) \tag{2.17}$$

(2.17) expresses a lower bound since instances of schema S may be generated by crossover of strings not in S.

Hence the expected number of survivors of a schema S subjected to crossover is,

$$n(S,t+1) >= n(S, t)(F(S)/F')(1-Pc\ d/(L-1))(1-P[S,t]) \tag{2.18}$$

We now extend Eqn (2.18) to include the effect of mutation. Survival of a schema will occur when all the defined bits survive, this probability is,

$$(1 - Pm)^{o(S)} \tag{2.19}$$

For small values of Pm this can be taken as,

$$1 - o(S) Pm \qquad (2.20)$$

Hence (2.18) can be extended to include mutation,

$$n(S,t+1) \geq n(S,t)(F(S)/F')\{(1-Pc\ d/(L-1))(1-P[S,t])-o(S)Pm\} \qquad (2.21)$$

From the form of (2.21) we conclude that schemata with above average fitness, with short defining lengths and low order specificity of fixed positions will receive exponentially increasing numbers in subsequent generations. Such schemata are of such importance in the GA that they are given the special name of 'building blocks'.

Whilst the building block hypothesis is generally reliable in most practical problems there are cases where this is not so. An example of this is the 'deceptive' type of problem where optimum values of some or all of the variables are isolated in the combinatorial space and remote from 'near' optimum values so that these latter tend to draw the GA away from the optimum values.

2.9. CODING CONTINUOUS-VALUED VARIABLES

If continuous functions are mapped onto a binary string then we need to consider the degree of precision achieved. If we map to a binary string of length L and attempt to represent a real continuous variable lying between an upper limit a and a lower limit b, then the degree of precision can be interpreted as the interval between two adjacent decoded values,

$$(a - b) / (2^L - 1) \qquad (2.22)$$

The degree of precision can be improved by increasing the string length L.

2.10. BIT ARRAY REPRESENTATIONS

A binary representation in two-dimensional form, called a 'bit-array', has been described by Schoenauer [17] and applied to optimal shape problems. Basically the complete design space, representing a cross-section, is partitioned into two subsets, material and void. A '1' indicates the presence of material at that position in the cross section and a ' 0' indicates no material. Fig.2.4 shows a tee-section described in this way.

When using bit-arrays the genetic operators are specially designed for operating in two-dimensions.

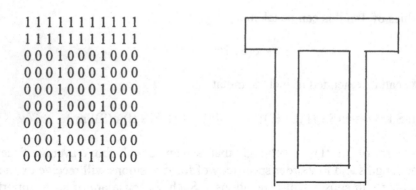

```
1 1 1 1 1 1 1 1 1 1
1 1 1 1 1 1 1 1 1 1
0 0 0 1 0 0 0 1 0 0 0
0 0 0 1 0 0 0 1 0 0 0
0 0 0 1 0 0 0 1 0 0 0
0 0 0 1 0 0 0 1 0 0 0
0 0 0 1 0 0 0 1 0 0 0
0 0 0 1 0 0 0 1 0 0 0
0 0 0 1 1 1 1 1 0 0 0
```

Fig.2.4: Bit-array representation of hollow tee-section

2.11. POPULATION CONTROL AND DIVERSITY

In many applications of the GA, the population in generation i will be replaced, more-or-less, completely by a new population in generation i+1. As one would expect, some of the individuals will be reinserted into the population unchanged since identical bits in corresponding positions in a string will not be changed by crossover. This may not be a good thing in that some good individuals may be lost in the process. In some versions of the GA the composition of the population is only partially changed so that selected good individuals can be retained.

We have seen that crossover in itself cannot introduce new genetic material, for example if a 1 bit exists at a specific location in the optimum string then if this bit is not present in any of the original population, it cannot be generated by crossover. It can however be generated by mutation. Mutation encourages diversity introducing new information by flipping bits. The mutation operator needs restrained control otherwise important information can be lost. Generally a high mutation rate (say 0.01) may be appropriate in the early stages of a GA run whereas a lower rate (say 0.001) may be more effective in the later stages. In some operator-adaptation rules, mutation is adapted to zero as in (2.13). We now list some of the more advanced devices used in applying the GA.

In 'crowding' a newly formed offspring replaces the existing individual most similar to itself - preventing too many similar 'crowds' being formed and preserving diversity. 'Speciation' encourages the formation of stable subpopulations or 'species' where members converge on several peaks ('niches') to prevent all congregating on a single peak. A device called 'sharing' achieves this by allocating a high weighting to strings close to an individual and a lower weighting to more distant strings. The fitness of each individual is then adjusted in accordance with the accumulated shares. 'Mating restrictions' - may disallow mating between similar individuals.

'Rank selection' - is claimed to inhibit premature convergence - the individuals in the population are ranked according to fitness and the expected value of each individual depends on its rank rather than its fitness. Ranking avoids giving the largest share of offspring to a small group of highly fit individuals. In the linear ranking method (Baker[19]),

$$\text{adjusted fitness} = \text{max} - (\text{max-min})*(n - R)/(n - 1) \quad (2.23)$$

where R = rank, 1<= Max <= 2. Baker recommended Max = 1.1. Ranking may slow-down selection pressure.

2.12. ELITISM

If the selection of individuals is based only on the principles described so far, then it is possible for the current best individual to have a smaller fitness than the best individual in the previous generation. It is tempting to try to avoid this by inserting the current best individual into the next generation. Unless special steps are taken, there is however no guarantee that the best individual will survive. Figure 5 shows what might happen if we introduce elitism into the GA. The result at generation 100 shows an improvement but thereafter the performance of the algorithm degrades. Effective use of the GA requires a combination of sensitive control settings along with freedom for the algorithm.

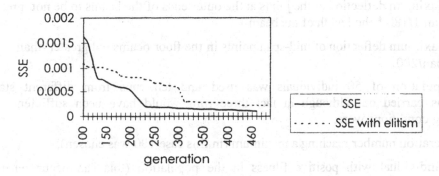

Fig.2.5: GA training with and without elitism (SSE = sum-squared error)

2.13. FURTHER CONSIDERATION OF STRUCTURAL DESIGN OPTIMIZATION

We consider the optimum (minimum weight) design of the multi-storey frame shown in Fig.2.6. An optimization study of this structure was published, Jenkins [18] so what follows is a brief summary of the work carried out. The structure is a plane rigid frame with the outer ends of the floor beams supported by hangers from a truss at roof level.

The applied loading of 20 kN/m^2 is uniformly distributed on floors 1, 2, .. 6. The structural design is described in terms of 9 design variables,

x_1 the beam cross-section at floors 1,2 and 3
x_2 the beam cross-section at floors 4,5,6 and 7
x_3 the column cross-section in storeys 1 and 2

x_4 the column cross-section in storeys 3 and 4
x_5 the column cross section in storeys 5,6 and 7
x_6 the cross-sections of the hangers
x_7 the cross-sections of the members of the truss
x_8 the length of the cantilever spans
x_9 the depth of the truss.

Design was carried out generally in accordance with the British Standard BS5950 with design constraints as follows,

(1) The hangers should be in tension. A small compression may be accepted.

(2) The tensile stress in the hangers must not be greater than 275 N/mm^2

(3) The bending stress in the beams must not be greater than 275 N/mm^2

(4) The combined bending and axial stress in the columns to satisfy BS5950

(5) The slenderness ratio of the columns to be not greater than 180

(6) The tensile stress in the truss members to be not greater than 275 N/mm^2

(7) The slenderness ratio of the struts in the truss to be not greater than 180

(8) Maximum deflection of the joints at the outer ends of the beams to be not greater than 1/180 * the length of the beam

(9) Maximum deflection of mid-span points in the floor beams not greater than span/200.

A population of 50 individuals was used and ten runs from different starting populations carried out although in fact three runs would have been sufficient. The terminating conditions were,

i)   generation number reaching a maximum (in this case 100 was chosen),

ii) no individual with positive fitness in the population (this can occur when the processing is dominated by the constraints and can be negotiated by easing some or all of the constraints in the early stages. This was not required with this structure.)

iii) A value of maximum fitness/average fitness approaching 1. This indicates that the population is tending to contain a large number of the same individual (a uniform population).

It is generally found desirable to set a maximum generation number such that termination is based on one of the other conditions. Condition iii) is the one most usually applying. As each generation was reached, the fitness of each individual was computed, each structure analysed using an 'exact' stiffness matrix method, the design constraints assessed and penalties applied as necessary.

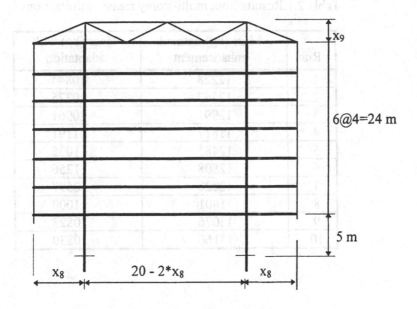

Fig.2.6.: Multi-storey frame for optimization study

Two sets of results are quoted in Table 1, the first is the mass of the structure (kg) with no enhancement of the GA and the second with the control adaptation described in Section 6 and also the space condensation heuristic which is to be described in Section 2.14. In handling the design constraints it has been found useful to observe the practical device of permitting a small tolerance, this would generally be the case in design. The program provides an option to select a specified tolerance on the constraints. A zero tolerance was imposed in this example.

An optimization study was carried out on the cable-stayed structure shown in Fig.2.7 (Jenkins [6]) using the genetic algorithm coupled with a plane-frame program for structural reanalysis. The structural optimization was concentrated on the 17 design variables shown in Fig.2.7. A binary string length of 4 was used for each variable. For a uniformly distributed load over the complete span, the design constraints were,

   (i)   maximum stress in cables
  (ii)   maximum combined stress (bending + axial) in the girder
 (iii) maximum combined stress in the tower
 (iv) maximum vertical displacement in the girder, and
  (v) maximum lateral displacement at the top of the tower.

To illustrate the suitability of the algorithm for design parametric studies, optimization was carried out for the alternative tower heights shown, for alternative cross-sectional properties of the towers and for a range of values of the ratio unit cost of cable material/unit cost of structural steel.

Table 2.1.Results from multi-storey frame optimization

Run	Mass (kg) without enhancement	Mass (kg) with adaptation
1	12228	10844
2	12567	10775
3	12991	10961
4	11817	11191
5	12485	11035
6	12808	11256
7	12020	10597
8	11801	11000
9	13076	10823
10	13166	10830

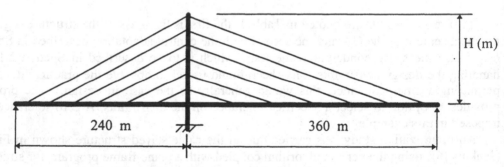

x_1 = spacing of cable attachments in 240 m span
x_2 = spacing of cable attachments in 360 m span
x_3 = spacing of cable attachments on tower
x_4 .. x_{15}= cross-sectional areas of 12 cables
x_{16} = cross-section of girder/deck
x_{17} = cross-section of tower, tower height H = 90 m (Case A), 120 m (Case B)

Fig.2.7: Cable-stayed bridge for optimization study

2.14. REDUCING THE COMBINATORIAL SPACE

In Section 2.2 we observed that if we use very long strings to represent a concatenation of a complete individual, then single point cross-over is unlikely to result in reasonably rapid progress. If we are to seek some form of string partitioning then it is very attractive to partition on the basis of the individual variables [4]. This would have clear advantages in the processing and would very conveniently allow us to use different string lengths for each variable if we desired, moreover we could introduce a process whereby the string lengths could change *during* the processing. So by adopting a variable-by-variable approach to the processing we then admit several attendant advantages. If, for example. we could establish during the processing that some of the parameter values being offered to the algorithm were much preferred as compared with other values, then we could eliminate the unpopular values by shortening the appropriate string and re-coding the remaining values. Shortening a string by one bit will result in the number of values offered being reduced by one-half. If we could be sure that the values retained contain the 'best' then we could reduce the combinatorial space, prevent the GA from venturing into some low-fitness areas and hopefully speed the operation of the algorithm. The details of the space condensation heuristic have been presented, Jenkins [15],[24] and will be summarised below.

In passing we should note that other authors [21] have proposed methods not requiring reductions in the search space, adopting a dynamic shift of emphasis in the space using 'stochastic string coding'. The total space is partitioned into 'regions' and comparative values are applied to the regions. The definitions of the regions are dynamic and change during the processing.

2.15. ANALYSIS OF PARAMETER VALUE SELECTION INTENSITIES

2.15.1. The basis of a condensation heuristic

In order to provide the heuristic with a basis on which to condense the search space, a record is established of the parameter values selected by the algorithm during the processing. For any typical variable, values associated with relatively high fitness individuals and those associated with relatively low fitness individuals are recorded. The record is maintained in a two-dimensional array (PSELS).

A fraction (δ_u) of the above-average individuals is recorded by adding 1 into the appropriate cell in the array for each coded parameter value selected within the fraction. Similarly 1 is subtracted from the appropriate cell for those parameter values selected for individuals within a fraction (δ_l) at the bottom end of the fitness range. The result is that values consistently associated with the best individuals will grow positively whereas values consistently associated with low fitness individuals will decrease. As the processing continues generation-by-generation we will expect to see trends in selection patterns. The structure of the record array is shown in Fig.2 8. We now look to see if we can identify and exploit these trends. At the end of processing in each generation we look for two features in the record in respect of each variable. Suppose we examine the record for variable x_i, identify the parameter value showing the maximum number of selections (MSEL) and compare this with a threshold value λ. If MSEL has reached the threshold then we proceed to test the distribution to assess whether MSEL shows a significant preference compared with the other parameter values in the set. To do this we sort the selection numbers from MSEL downwards to establish the selection frequency distribution. If this is dominated by MSEL then we may have sufficient confidence in the processing to condense the list of values offered for x_i by reducing the chromosome length by one and thus halving the number of values provided. The combinatorial space is immediately divided by two whenever a condensation takes place. With a reasonably cautious operation of the heuristic it is likely that the optimum value of the variable x_i will be retained in the condensed list.

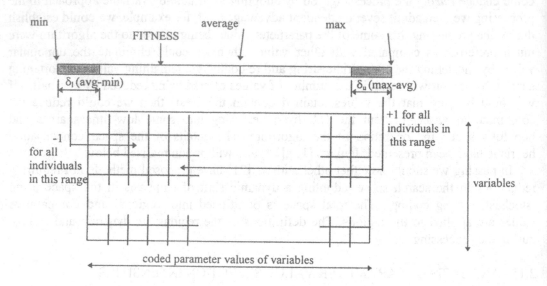

Fig.2.8.: Structure of parameter value selection record (PSELS)

A suitable classifier for use in an assessment of the selection distribution is the general regression neural network [22] but before proceeding to this we consider two further adaptations which are made on-line.

2.15.2. Adaptation of the upper and lower selection fractions

In the early generations of a GA run, all fitnesses can be assumed to be low relative to what is to come. It is therefore appropriate to record a substantial number of the lower fitness individuals and only a few of the individuals with higher fitnesses. In the late stages of the run the reverse is true and it would be appropriate to record most of the higher fitness individuals and few of those with lower fitnesses. This can be achieved using the following adaptations,

$$\delta_u = \delta_{uo} * (avg/max)^2 \qquad (2.24)$$

$$\delta_l = \delta_{lo} * (1 - (avg/max)^2) \qquad (2.25)$$

where, δ_{uo} and δ_{lo} are initial values of the fractions chosen by the user, avg and max are the current average and maximum fitnesses in the population. It should be noted that the ratio (avg/max) rises quickly towards 0.9 and higher, indeed a very high value of this ratio (eg 0.9999) is used as a terminating condition for the algorithm.

2.15.3. Adaptation of the threshold λ

Having decided what would be a suitable value of the threshold λ at the initial stage of the algorithm, we need to observe that, for each variable, as the chromosome length reduces we shall need to increase λ in order to maintain the same level of significance since the algorithm is now selecting from a reduced list. The updating rule for λ is,

$$\lambda = \lambda_0 * (2^{lco[i]}) / (2^{lc[i]}) \tag{2.26}$$

where, λ_0 = initial value of λ

$lco[i]$ = initial value of chromosome length for variable i

and, $lc[i]$ = outgoing chromosome length of variable i.

As an example of the operation of (2.26), if the initial value of λ was chosen as 50, then successive values at condensations would be, 100, 200, 400 .....

2.15.4. Classification by general regression neural network

The classification will be simplified if we sort the parameter value selection numbers into decreasing order, as depicted at (a) in Fig. 2.9.

The corresponding decoded parameter values will be arranged in the same order. Suppose the result is a distribution similar to that shown at (a) in Fig.2.9 with MSEL being the parameter value showing the largest number of selections recorded at the current stage. The ordinates in the diagram are the numbers of selections made for each of the coded values for the current variable. The output of the GRNN classifier is binary $[y_1, y_2]$ where $y_1 + y_2 = 1$. A distribution as at (b) will produce an output $[1, 0]$ and depicts a situation where one parameter value is selected uniformly. An output $[0, 1]$ will result if the distribution is as shown at (c) in which all parameter values are selected equally. Distributions (b) and (c) form the basis of training data for the GRNN. A triangular distribution will be indicated if the GRNN output is $[0.5, 0.5]$, as at Fig.2.9 (d). We observe that the extreme distributions, (b) and (c), are independent of the abscissae in the data presented to the classifier. Now we can improve the performance of the classifier if we introduce the influence of the positional relationships of the decoded parameter values. Close decoded values should have small differences in absicssae, since they are mutually supportive in the classification, whereas distant decoded values should be spaced further apart. Now if we are to represent real decoded values of the variables and hence a non-uniform interval for the absicssae, then the form of the training functions must be changed to suit. It is clear that we should use differences in the decoded values of the variables as abscissae (X) when plotting a distribution. The following expression has been used for representing the abscissae in the distribution,

$$Xj = \sum_{i=1}^{j} ABS(x_{i+1} - x_1); \quad j = 2,3 \ldots nv \tag{2.27}$$

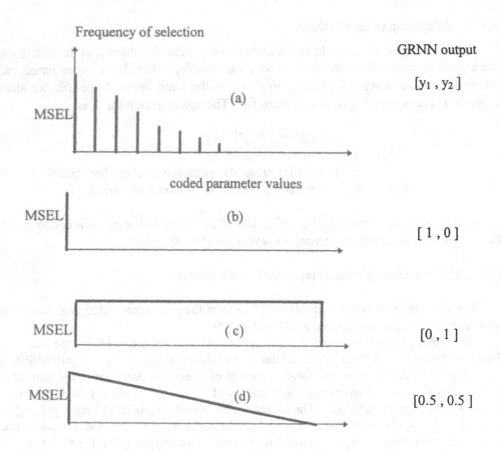

Fig.2.9.: Distributions of parameter value selections
with corresponding GRNN outputs

where, X_1 = 0;
      x_1 = decoded parameter value corresponding to the maximum
            selection,
      x_i = decoded parameter value corresponding to the i th item in the sorted list
            of selection numbers.
     nv = number of parameter values available for selection (= 2^{lc}).

    The values X are normalized by division by the maximum value (X_{nv}) and the resulting abscissae then range from 0 to 1. Similarly the ordinates are normalized by division by MSEL. It follows that parameter values numerically 'near' to the most popular value will have small abscissae near the origin. If the numbers of selections recorded for these 'near' values are similar to that of the most popular value, then the shape of the training function must model this. A suitable shape is the one given by (2.28) and shown as u_1 in Fig. 10.

$$u_1 = EXP(-100*X[i])^3) \tag{2.28}$$

Consider now the shape of the training function u_2; here we need to provide for the situation where a cluster of values remote from the maximum might trigger the heuristic. This action can be deterred by using a function such as u_2, (2.29), shown in Fig. 2.10.

$$u_2 = 1 - EXP(20 * (X[i] - 1)) \tag{2.29}$$

The precise form of these functions is less important than the shape.

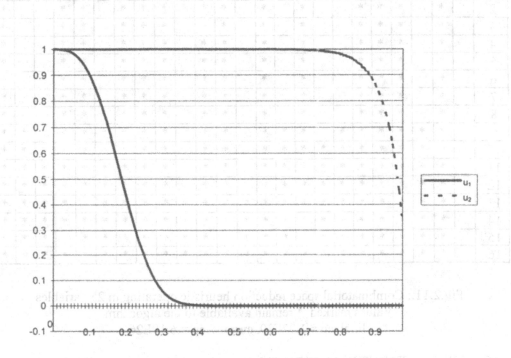

Fig.2.10.: Training vectors for GRNN classifier

Some work has been carried out in an attempt to 'calibrate' the heuristic, [13,15], using the 'ideal' distribution of Fig.2.9 (b). Using results obtained from this investigation it is possible to exert some control over the quality of behaviour of the heuristic by selecting 'high' or 'low' quality processing. The high quality option requires the GRNN output to lie nearer to the 'ideal' distribution whereas the low quality option allows a greater margin of difference. Reference [15] should be consulted for more details of this.

The action of the heuristic is illustrated in Fig.2.11 which is a 'snapshot' of the process at a particular generation in the processing and shows the 'coded' values remaining available to the algorithm. It should be noted that the decoded real values are re-sorted each time a condensation is invoked. Initial and current combinatorial products are also shown in

Fig.2.11. When the chromosome length for a variable is reduced to 1, then this can hold two coded values of the variables (0 and 1). Since the GA cannot be effective in these circumstances, a simple competition is used to choose the best of the two. Thus, in suitable circumstances, the combinatorial product can be reduced to 1 resulting in a single unique solution.

Coded values	2	3	4	5	6	7	8	9	10	11	12	13	14	15	16	17	18	19	20	21	22	23	24	25
1	*		*	*	*			*		*		*	*		*	*	*		*	*			*	*
2	*		*	*	*			*		*		*	*		*	*	*		*	*			*	*
3	*		*	*	*			*		*		*	*		*	*	*		*	*			*	*
4	*		*	*	*			*		*		*	*		*	*	*		*	*			*	*
5	*		*	*	*			*		*		*	*		*	*	*		*	*			*	*
6	*		*	*	*			*		*		*	*'		*	*	*		*	*			*	*
7	*		*	*	*			*		*		*	*		*	*	*		*	*			*	*
8	*		*	*	*			*		*		*	*		*	*	*		*	*			*	*
9	*	*	*	*	*	*	*	*	*	*		*	*		*	*	*	*	*	*	*	*	*	*
10	*	*	*	*	*	*	*	*	*	*		*	*		*	*	*	*	*	*	*	*	*	*
11	*	*	*	*	*	*	*	*	*	*		*	*		*	*	*	*	*	*	*	*	*	*
12	*	*	*	*	*	*	*	*	*	*		*	*		*	*	*	*	*	*	*	*	*	*
13	*	*	*	*	*	*	*	*	*	*	*	*	*	*	*	*	*	*	*	*	*	*	*	*
14	*	*	*	*	*	*	*	*	*	*	*	*	*	*	*	*	*	*	*	*	*	*	*	*
15	*	*	*	*	*	*	*	*	*	*	*	*	*	*	*	*	*	*	*	*	*	*	*	*
16	*	*	*	*	*	*	*	*	*	*	*	*	*	*	*	*	*	*	*	*	*	*	*	*

Fig.2.11.: Combinatorial space reduction heuristic operating in 25 variables.
Values marked * remain available to the algorithm.
Iitial cp = 1.27E30, current cp = 6.19E26

2.16. TERMINATING THE ALGORITHM

With the initial population being based on random selection of individuals and the GA controls being stochastic in nature, the behaviour of the GA in any run is, in practice, unpredictable. It would be usual to run the GA a number of times from different initial populations and compare results, five runs is often sufficient. As a terminating condition, a maximum can be placed on the number of generations processed. This is unsatisfactory as a sole terminating condition since in a few more generations a significant improvement may be achieved. A better procedure is to impose a number of terminating conditions as follows,

(1) gen = maxgen (say 2000) - the purpose being to impose a number of generations which the GA is unlikely to reach having terminated before this.

(2) combinatorial product = 1; (this is achieved in some circumstances)

(3) average fitness / maximum fitness > 0.99999 (this is the most likely terminating condition indicating that the population is composed uniformly of similar individuals).

(4) no individual with positive fitness in the population. (This condition can occur in severely constrained circumstances and can be handled by easing some of the constraints in the early stages of processing).

2.17. THE ESTIMATION OF PARTIAL STRING FITNESSES

There is considerable evidence of the effectiveness of the GA when the processing is carried out on partitioned strings, each partition corresponding to one variable in the objective function, however it would be more satisfactory if we were to explore the technique at slightly greater depth to try to establish some justification. The parameter selection intensity record (PSELS) described above can be enhanced by the addition of another record based on accumulated fitnesses (SIGFITS). Some preliminary studies had shown that a high level of accumulated fitnesses accompanied by a relatively low level of actual selections indicated a 'good' value for the variable in question.

In updating the record, the fitnesses allocated to each variable are normalized by division by the average fitness for the current generation. The normalized fitnesses are then aggregated in an array SIGFITS[i,j] and a ratio (sr = selection ratio) is formed as follows,

$$sr = SIGFITS[i,j]/PSELS[i,j] \qquad (2.30)$$

The ratio sr is used to apply factors to the individual fitnesses as follows,

$$PFIT[i] = FIT[i] * ((1 - pff) + pff * sr) \qquad (2.31)$$

where PFIT[i] and FIT[i] are the scaled partial fitness and raw fitness of individual i, pff is a user-specified partial fitness factor and sr the ratio defined in eqn (2.30). The partial fitness factor is used to weaken or strengthen the effect of sr. If pff = 1, then the partial fitness ratio sr is fully applied whereas if pff = 0, then the adjustment is withheld.

Some results published, Jenkins [25], showed that the application of the partial fitness scaling improved the rate of convergence.

2.18. MULTI-OBJECTIVE GA – PARETO OPTIMALITY

When it is necessary to specify more than one criterion in an optimization then the problem is termed 'multi-criterion' or 'multi-objective'. Suppose we are engaged in an optimization in which there are m objective functions $f_i(x)$ (0<=i<=m). A point x^* is Pareto optimal if no objective can be improved without worsening at least one other objective. In other words, there are no points better on *all* criteria. The Pareto optimum generally results in a set of solutions called 'non-inferior' or 'non-dominated' [27].

It is convenient to change the multi-criterion optimization problem to a scalar optimization problem by aggregating weighted individual objectives as follows,

$$f(x) = \Sigma \; w_i \, f_i \, (x) \quad (1 <= i <= m) \tag{2.32}$$

where we define weighting factors w_1, w_2 ... w_m to be applied to each criterion. The weights function as probabilities, i.e.,

$$0 <= w_i <= 1, \quad \text{and,} \; \Sigma w_i = 1 \tag{2.33}$$

Fig.2.12 shows two possible scenarios with two objective functions. In (a) the multi-objective solution is obvious and coincides with the separate optima for both function, in (b) no immediately identified multicriterion optimum is obvious but in general the best results will lie to the left and downwards. Different pairs of weights satisfying (2.32) will produce different solutions, eg A, B, C, in Fig.2.12. The other points shown are 'dominated'.

The effect of the weighting coefficients is not straight-forward and requires that the separate objective functions be normalized otherwise a function having a comparatively large value can dominate the optimization procedure. When substituting in eqn (2.32), each objective function is divided by its minimum value. The multi-objective optimization process is then applied to the weighted aggregate of objective functions as represented by (2.32).

Several authors have used the GA in multi-objective optimization. Mention is made in the references to the work of Koumousis [26] and Cheng [29]. In optimizing reinforced concrete beams Koumousis [26] used a weighting method with scaling for,

(a) minimum weight, to counter a tendency towards very small bar diameters and the resulting increase in anchorage lengths required,

(b) uniformity, to introduce a reasonable level of uniformity in bar diameters whilst still achieving a 'near-optimum' structure.

In the work described by Cheng and Li [29] the usual GA operators of crossover and mutation were augmented by two additional operators,

(i) niche: this constrains individuals to share available resources and maintain diversity in the population. This is particularly appropriate since the solution of a Pareto GA is a region rather than a point.

(ii) A Pareto set filter: this pools non-dominated points at each generation and reduces the effects of genetic drift.

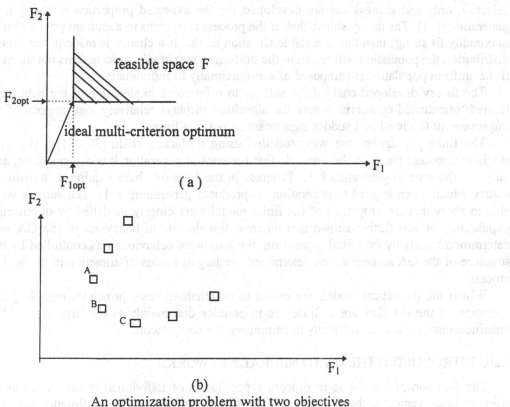

(b)
An optimization problem with two objectives
The Pareto (non-dominated points) are labelled A, B, C

Fig.2.12.: Multi-objective optimization

2.19. MATHEMATICAL MODELLING OF THE GENETIC ALGORITHM

In Section 2.8 we considered the importance of schemata in the operation of the GA and evaluated probabilities of schema survival during the operations of population progression. We observed that certain schemata, those with above average fitness, short defining lengths and low order specificity of fixed positions, would receive exponentially increasing numbers in subsequent generations. This does not however provide any direct information regarding the expected fitness distributions or the rate at which convergence might take place. In order to do this we need to look at mathematical modelling of the process. We look briefly at the mathematical models devised by Vose and Liepens (1991) as described by Mitchell [3].

The general model is based on a simple GA. Fitness is calculated for all individuals, two parents chosen and crossover carried out at a random site at probability p_c forming two offspring. Randomly select one and discard the other, mutate each bit at probability p_m and repeat the process until the population is full. Initially the model is constructed representing selection only and expressions are developed for the expected proportion of string k in generation (t+1). The theory shows that if the process converges to a uniform population of maximally fit strings then this is a stable situation so that if a change is made in the fitness distribution the population will return to the uniformly maximally fit state. This not the case if the uniform population is composed of sub-maximally fit individuals.

The theory developed enabled the authors to offer some explanation of the behaviour termed 'punctuated equilibria' where the algorithm displays relatively long periods of no improvement followed by a sudden significant increase in fitness.

The finite population case was modelled using a Markov chain [30],[31]. This is a stochastic process based on the principle that the state at generation k is dependent on, and only on, the state at generation k-1. Progress in the Markov chain requires a 'transition' matrix which when applied to generation t produces generation t+1. The authors were able to show that the properties of the finite model were closely modelled by the infinite population. It was further shown that whereas the short-term behaviour of the GA was determined largely by the initial population, the long-term behaviour was controlled by the structure of the GA surface which determined the largest basins of attraction in the search process.

Whilst the theoretical models are useful in establishing basic principles regarding the operation of the GA they are of little use in practice due mainly to the large size of the transition matrices and the difficulty in computing the components.

2.20. INTRODUCING THE GA TO NEURAL NETWORKS

The function of the GA is to process a population of individual 'solutions' so as to effect an improvement in the direction of an 'optimum' or 'best' situation. Broadly there are three areas in which the GA can be applied to neural networks,

  i)   improving the network topology,
  ii) finding the best values for the network controls, and
  iii) training the weights.

The first two applications are relatively straight-forward and considerable progress has been made, however the third application is more difficult and presents us with more problems. Since we are to concentrate on an application in weight training we need to look at one of the problems arising. Some authors, Braun/Zagorski [35], have adopted hybrid approaches when applying evolutionary methods to neural network training.

2.20.1. The 'permutations' problem in weight training by neural network

This problem arises from the population-based feature of the GA and the recombination of individuals as the algorithm proceeds from one generation to the next. If we consider a neural net with input, output layers and one hidden layer then we see that whilst the relative positions of the nodes in the input and output layers are fixed, by definition, the relative positions of the nodes in the hidden layer are arbitrary and these nodes can be repositioned without affecting the functionality of the network. The number of possible positional relationships of the nodes in a hidden layer of h nodes is h ! .

At any stage in the processing each individual could be identified with a different permutation of the hidden unit positions from the other individuals. Recombination is then in danger of producing confusion by duplicating the information from some nodes and omitting information from certain others. A number of suggestions have been made to solve this problem, It has been suggested that the situation can be countered by high mutation rates, or by specially contrived cross-over operators which rely on identifying the functional relationships of the units.

The permutations problem could be removed if we were to specify fixed positional relationships of the hidden nodes as we do with the input and output nodes. If we recall the form of string coding we explored in Section 2.2 and observe that with this form of coding we operate on the complete individual in partial strings, each sub-string representing a specified variable (weight) then we see that recombination takes place on the basis of individual weights rather than the complete set. It can be concluded that the permutations problem will not occur with this type of recombination.

Some authors have concluded that the use of the GA should be restricted to a coarse level of weight training then handing over the fine tuning to a gradient-based method. This is unnecessary, we shall see that the GA can readily perform all the weight training through to a desired degree of accuracy.

2.20.2 Neural network training by ga with 'shift'

Some important changes are needed in the design of a GA to operate in weight training for a neural network. The GA selects coded values representing the variables, weights in this case. The set of values from which the weights are selected is finite and discrete however the GA needs freedom to select from an unlimited range of real values. In what we are to do, this requirement will be met by allowing the centre of the range of decoded values to 'shift' under pressure from the algorithm. The coded values will remain constant but the decoding rule will implement this shift.

2.20.2.1. Coding the weights

A chromosome length lc is chosen defining the length of the binary string for each weight. The coded values are decoded to discrete +/- integer values with zero at the centre. The general rule for decoding n values is,

$$-(n/2-1), -(n/2-2), \ldots\ldots 0, 1, 2, \ldots\ldots (n/2) \tag{2.34}$$

where $n = 2^{lc}$

For example if lc = 10 for each weight, the discrete values available to the algorithm will be,

$$-511, -510, \ldots\ldots\ldots 0, 1, 2, \ldots\ldots 512$$

These values will not be changed during the processing but the result of using them will change by adding a varying term called the 'shift' to each decoded value.   It will be found that the process works well even with very short strings, lc = 3 or 4, for example.

2.20.2.2 The concept of 'shift'

Suppose we have a chromosome length of 3 and start the processing with decoded values (-3, -2, -1, 0, 1, 2, 3, 4) and suppose that at some point in time, generation number gen , we examine the best set of weights and note that a particular weight w has a value 3. Now the value 3 lies near the positive end of the range and this might suggest that the algorithm is seeking to move further in this direction in the selection process for this weight.   If this is so we can assist the algorithm by identifying the 'best' individual in the population and 'shifting' the range of values so that the current best value of the weight w lies at the centre of the effective range.   This is achieved by applying a zero coded value to this weight. In order to keep a record of the state of shift, we initialize an array to maintain a cumulative value of shift for each weight in the network.   The shift process is invoked at each improved generation, after an initial 'settling down' threshold, and the current values of the weights for the best individual are 'added' into the array.   Thus the updating rule for the array 'shift' is,

$$shift[\dots]_{gen+1} \;=\; shift[\dots]_{gen} \;+\; w[\dots]_{gen} \tag{2.35}$$

The array contains a value for each individual weight.

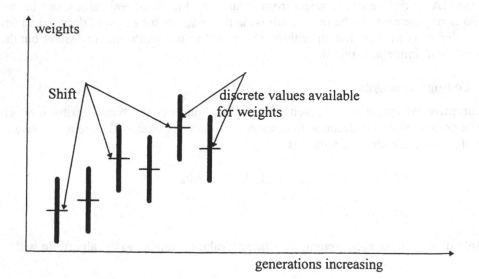

Fig.2.13.: Progressive shift in discrete values available to the algorithm

2.20.2.3. The fitness function

The fitness function for a set of weights must be related to the current error exhibited by the set. Suppose we have m output nodes and p training pairs in our network; we shall use the average sum squared error (SSE) as an error metric obtained when we run the net with the current set of weights and compare the results o[i] with the corresponding outputs in the training set t[i],

$$SSE = 1/p \sum_p \sum_m (t_i - o_i)^2 \qquad (2.36)$$

The GA will seek to increase fitness so, In order to reduce the error we set,

$$fitness = 1 - SSE \qquad (2.37)$$

2.20.2.4. Adaptation of controls

It can be useful to introduce an on-line adaptation of the slope of the transfer function, here we use the Sigmoid,

$$y = 1 / (1 + e^{-k(net)}) \qquad (2.38)$$

where, $k(net) = k * \sum (weighted\ inputs)$

If k(net) in (2.38) is > about 10, then the Sigmoid acts as a 'hard' limiter and excludes some individuals with relatively good fitness. On the other hand, if k(net) is low then the transfer will admit many weak individuals. It is found that the GA driven NN is sensitive to the value of k used. Some tentative investigations have suggested an adaptation of the form,

$$k = a(1 + (gen/maxgen)^2) * 1/(1+n) \qquad (2.39)$$

where n = the number of input nodes and a is of the order of 40.

2.20.2.5. Network topology

We shall illustrate network training using an example from Jenkins [34], in which a network is trained to provide a structural analysis for a generalized structural grillage of the type sometimes used in highway bridge design. The network topology consists of an input layer with 5 nodes, a hidden layer with 9 nodes and an output layer having 4 nodes.
The design data presented to the input layer were as follows,

x_1 span (L)
x_2 number of longitudinal members in the grillage (N_x)
x_3 number of transverse members in the grillage (N_y)
x_4 second moment of area of all members (I)
x_5 torsional constant of all members (J)

The output from the net consisted of,

   o_1 maximum bending moment in the longitudinal members
   o_2 maximum bending moment in the transverse members
   o_3 maximum twisting moment
   o_4 transverse deflection at centre of grillage.

The applied load was a uniformly distributed load over the whole plan area of the grillage.

2.20.2.6 Training data

The network was trained on 13 sets of data and the trained network tested on 10 unseen sets.   The training data is listed in rows 1 .. 13 inclusive in Table 2 and the test data in rows 14 ... 23. Results are quoted from 'exact' structural analyses and by neural networks trained by back-propagation and by GA. The results are comparable and satisfactory from the two neural network training methods. A trace on some sample weights in the GA trained network are shown in Fig. 14.   This shows that the shift process is driven in a clear direction by pressure from the algorithm. Quite short chromosome lengths (3 or 4) are found to be satisfactory. The genetic operators, selection/ crossover/ mutation, are applied to partial strings separately for each variable weight. It was interesting to note that the BP algorithm made more rapid progress early in the processing whereas the GA appeared to be more efficient in the later stages. The situation is however insufficiently general to allow conclusions to be drawn.

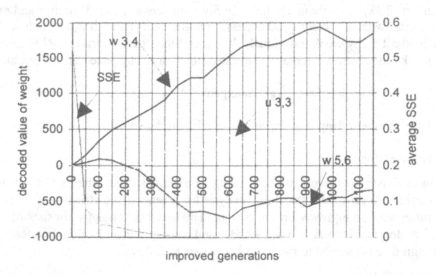

Fig.2.14.: Grillage analysis network trained by GA using shift.

We note that the GA searches the 'entire' combinatorial space whereas BP follows a gradient path through the space. This suggests that the GA might conduct a more effective search in the early stages of processing because of its wide-ranging action. This could act as a handicap in the later stages by taking and retaking the algorithm into weak areas of the combinatorial space. At this stage the BP algorithm might be expected to be more efficient. Further work is needed before this principle could be established. It has not been particularly evident in the work conducted so far.

Table 2.2: Neural network analysis of structural grillage
using back-propagation (BP) and genetic algorithm (GA).
Overall SSE on test data; BP=0.00048, GA=0.0084, Units kNm, mm.

#	L	Nx	Ny	I	J	\multicolumn{4}{c}{exact}	\multicolumn{4}{c}{BP}	\multicolumn{4}{c}{GA}
						Mx	My	Mxy	D	Mx	My	Mxy	D	Mx	My	Mxy	D
1	8	4	4	0.02	0.02	155.1	22.3	6.14	1.5	149.1	22.5	6.14	1.6	148.1	22.1	6.09	1.9
2	20	10	10	0.33	0.08	395.5	18.5	1.04	1.6	393.7	19.0	1.07	1.7	397.6	18.0	1.13	1.2
3	8	7	10	0.11	0.05	93.7	7.8	1.42	0.2	95.0	8.4	1.40	0.2	97.4	9.2	1.56	0.3
4	20	7	10	0.11	0.05	565.0	24.9	2.52	7.1	555.5	25.8	2.48	7.2	562.4	25.2	2.51	7.2
5	14	4	10	0.11	0.05	486.3	28.4	3.13	3.0	486.6	29.0	3.09	3.0	486.0	28.9	3.21	2.5
6	14	10	10	0.11	0.05	194.8	11.9	1.42	1.2	190.9	12.6	1.36	1.2	195.5	11.6	1.43	1.2
7	14	7	4	0.11	0.05	256.1	36.9	3.48	1.4	262.8	37.1	3.37	1.5	256.1	37.1	3.50	1.9
8	14	7	16	0.11	0.05	279.5	11.0	1.48	1.7	284.1	10.7	1.34	1.9	279.8	11.7	1.51	1.6
9	14	7	10	0.02	0.05	278.2	11.8	5.79	9.4	287.8	11.6	5.73	9.2	281.0	11.4	5.73	9.4
10	14	7	10	0.33	0.05	278.2	19.2	0.81	0.6	283.5	19.1	0.81	0.5	282.5	19.3	0.60	0.2
11	14	7	10	0.11	0.02	278.2	19.1	0.95	1.7	281.7	18.7	1.13	1.5	283.4	19.1	1.00	1.7
12	14	7	10	0.11	0.08	278.3	16.5	2.94	1.7	268.6	16.3	3.08	1.8	278.3	15.7	2.84	2.1
13	14	7	10	0.11	0.05	278.3	17.7	2.07	1.7	280.1	17.8	2.00	1.7	275.0	16.5	2.09	1.8
SSE										.0000	.0000	.0000	.0000	.0000	.0000	.0000	.0000
14	14	10	8	0.04	0.06	193.4	10.7	3.29	3.2	190.6	12.5	3.47	3.0	171.6	12.6	3.80	4.1
15	10	8	12	0.02	0.04	125.3	5.2	3.26	2.2	122.7	6.4	2.45	1.6	113.8	7.2	2.50	1.6
16	18	10	4	0.24	0.07	291.8	37.2	1.84	1.2	200.3	31.4	1.67	1.2	363.3	25.7	1.64	1.4
17	10	4	16	0.17	0.08	251.1	11.8	1.81	0.5	202.5	11.2	1.80	0.4	193.1	11.7	1.12	0.3
18	8	7	6	0.11	0.03	92.0	12.2	1.34	0.2	88.3	12.7	1.27	0.2	93.0	18.1	1.77	0.3
19	18	9	12	0.02	0.07	357.8	9.7	5.69	20.0	377.4	9.6	5.10	16.4	391.9	12.7	4.93	12.7
20	12	7	6	0.24	0.08	202.2	23.9	2.11	0.4	187.0	23.4	2.18	0.3	193.9	23.8	1.19	0.2
21	14	4	14	0.11	0.06	488.4	19.6	2.83	3.0	445.9	17.3	2.70	3.3	455.3	19.0	2.95	2.7
22	16	5	10	0.04	0.02	507.4	29.0	3.23	11.2	574.0	23.1	4.67	14.1	580.4	19.3	4.20	16.3
23	20	6	14	0.07	0.05	663.1	19.9	3.33	13.1	609.9	15.1	3.75	17.5	613.4	16.4	3.76	12.9
SSE										.0014	.0008	.0017	.0009	.0022	.0024	.0021	.0017

2.21. PARALLEL GENETIC ALGORITHMS

Apart from the implicit parallelism of the GA, considerable attention has been directed to parallelism in hardware. A useful review of parallel implementations of the GA has been provided by CF Reeves [16]. The evaluation of individual fitnesses in parallel, using one processor for each individual is clearly an attractive approach as a counter to the high processing cost of a population-based method. There are communication overheads to consider. Selective allocation to parallel processors on the basis of sub-populations has been tried.

2.22. EXAMPLE OPTIMIZATION STUDY

We have formulated the GA in the context of optimum structural design and introduced illustrative examples as we progressed. It would be useful to spend a little more time considering the setting-up of a structural engineering design for optimization by GA so now we concentrate on a particular application of the GA. We consider a generalized structural grillage, similar to that mentioned in Point 2.20.2.5, and proceed stage-by-stage through the optimization process.

2.22.1. DEFINITION OF THE STRUCTURE

A plan view of the grillage is shown in Fig.2.15. The grillage spans in the X direction being simply-supported at Y=0 and Y=L (m). This type of structure is frequently used as a conceptual representation of a highway bridge in steel or concrete. The longitudinal and transverse members may be real or they may represent the cross-sectional properties of a reinforced concrete slab bridge. The grillage is rectangular in plan with N_x longitudinal members and N_y transverse members. The members are rigidly connected at their intersections so the connections transmit bending, shear and twisting stress resultants.

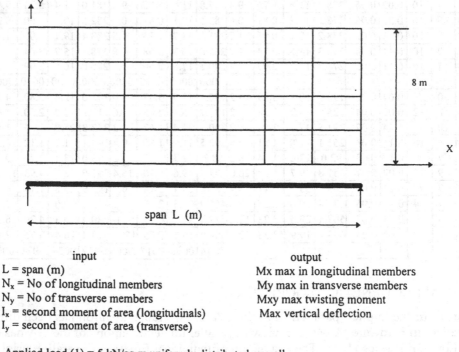

input
L = span (m)
N_x = No of longitudinal members
N_y = No of transverse members
I_x = second moment of area (longitudinals)
I_y = second moment of area (transverse)

output
Mx max in longitudinal members
My max in transverse members
Mxy max twisting moment
Max vertical deflection

Applied load (1) = 5 kN/sq m uniformly distributed overall
Applied load (2) = 120 kN knife edge load at mid-span.

Fig.2.15.: Layout of generalized structural grillage with input/output variables

2.22.2. DESIGN VARIABLES

In order to keep the treatment simple, we use five input design variables only: the span L (m), the number of longitudinal members N_x, the number of transverse members N_y, and the second moments of area of the longitudinal members I_l and the transverse members I_t. The torsional properties of the cross-sections are based on the second moments of area assuming the cross-sections to be solid rectangular (concrete).

Table 2.3: Discrete variables for grillage design

#	span L (m)	N_x	N_y	$I_l(m^4)$	$I_t(m^4)$
1	8	3	3	0.0054	0.0054
2	10	4	3	0.0074	0.0074
3	12	5	5	0.0100	0.0100
4	14	6	5	0.0171	0.0171
5	16	7	7	0.0273	0.0273
6	18	8	7	0.0417	0.0417
7	20	9	9	0.0610	0.0610
8	22	10	9	0.0860	0.0860

Again, in order to keep things simple and compact, we provide only eight values for each design variable as shown in Table 2.3. All designs will be composed of combinations of these five sets of eight values. In the GA these design data will be represented by binary strings each three bits in length thus providing 8 (2^3) discrete values as required. These strings will not be concatenated but handled separately in the GA.

2.22.3. Applied loading

Two types of load are considered in the design, (1) a uniformly distributed load of 5 kN/m^2 over the whole plan area of the grillage and (2) a knife edge load of 120 kN over the whole width of the grillage at mid-span. Three load factors (LF1, LF2 and LF3) are introduced and the design is to be carried out for two load cases (A) and (B) as follows,

(A) LF1 * (1) + self weight

(B) LF2 * (1) + LF3 * (2) + self weight.

2.22.4. Design constraints

For simplification only three design constraints were imposed, the maximum deflection to be not greater than span/500, the maximum bending stress to be not greater than 30 N/mm^2 in both longitudinal and transverse members. Penalty functions were devised according to the method outlined in Section 2.4.

2.22.5. Structural reanalysis

The GA processing requires a reanalysis of the structure whenever a change is made in the design variables. The subject of this design is a regular structure with longitudinal and transverse members arranged in a rectangular grid. This is an ideal situation in which to use a trained neural net to carry out the structural reanalysis.

2.22.6. Neural network architecture

A simple back-propagation network was used with input layer (5 nodes), output layer (4 nodes) and a single hidden layer. After a little experimentation it was decided to adopt 7 nodes in the hidden layer. The input data is normalized and the output mapped back to real numbers. This is a standard procedure in the author's network [33],[34]. The network uses a Sigmoid transfer function with back-propagation and adaptive gain.

2.22.7. Preparation of the training data

Training data were prepared by combining discrete data values for the five design variables using the 'hypercube' principle. First the 'corners' of the hypercube produced 32 sets (2^5) then the mid-faces of the hypercube (10 sets), then the centre of the hypercube (1 set). This made a total of 43 sets and a further 7 data sets were added representing combinations at 1/3, 2/3 points. Thus the complete training data was comprised of 50 sets. A further ten sets of data were prepared by random combinations, to be used in testing the trained network. The training data is listed in Table 2.4.

2.22.8. Training the net

The network was first trained simultaneously for all for outputs. The results were barely satisfactory so it was decided to train the net separately for each output, this produced satisfactory results. The net was therefore trained eight times, four outputs for design load (1) and four outputs for design load (2). Thus eight files of weights and other data were prepared for input to the GA optimizing part of the project. In all cases training was carried out to a target sum squared error of 0.0001. When applied to the test data the sum squared errors varied from a maximum of 0.0093 to a minimum of 0.00025. These error levels were regarded as satisfactory and the weights were then used to carry out the reanalysis during the operation of the GA.

2.22.9. The objective function

Optimization was carried out using as token objective the volume of material used in the design.

Table 2.4: Neural network training data for grillage analysis.
Combinations of coded data values from Table 2.4

#	L	N_x	N_y	I_l	I_t	#	L	N_x	N_y	I_l	I_t
1	1	1	1	1	1	26	8	8	1	1	8
2	1	1	1	1	8	27	8	8	1	8	1
3	1	1	1	8	1	28	8	8	1	8	8
4	1	1	1	8	8	29	8	8	8	1	1
5	1	1	8	1	1	30	8	8	8	1	8
6	1	1	8	1	8	31	8	8	8	8	1
7	1	1	8	8	1	32	8	8	8	8	8
8	1	1	8	8	8	33	1	4	4	4	4
9	1	8	1	1	1	34	8	4	4	4	4
10	1	8	1	1	8	35	4	1	4	4	4
11	1	8	1	8	1	36	4	8	4	4	4
12	1	8	1	8	8	37	4	4	1	4	4
13	1	8	8	1	1	38	4	4	8	4	4
14	1	8	8	1	8	39	4	4	4	1	4
15	1	8	8	8	1	40	4	4	4	8	4
16	1	8	8	8	8	41	4	4	4	4	1
17	8	1	1	1	1	42	4	4	4	4	8
18	8	1	1	1	8	43	4	4	4	4	4
19	8	1	1	8	1	44	5	5	5	5	5
20	8	1	1	8	8	45	2	2	2	2	2
21	8	1	8	1	1	46	7	7	7	7	7
22	8	1	8	1	8	47	2	2	2	7	7
23	8	1	8	8	1	48	7	7	7	2	2
24	8	1	8	8	8	49	3	3	6	6	6
25	8	8	1	1	1	50	2	6	6	6	2

2.22.10. Running the GA

The GA was used to optimize designs at fixed values of span ranging from 10 to 22 metres thus the optimization was confined to four design variables (N_x, N_y, I_l, and I_t). The values of load factors (LF1, LF2, LF3) used were 4,2 and 3 respectively. A population of 50 individuals was adopted in all the processing. Some early experimentation indicated that the GA would converge in a small number of generations so a maximum generation number of 100 was adopted.

The GA was run with adaptive crossover and mutation operators and with the space condensation heuristic operating. Initial values of controls were set as follows,

GRNN quality = low
pco = 1.0
pmo = 0.005
γ_0 = 0
λ_0 = 50
δ_{uo} = 0.1
δ_{lo} = 0.5

2.22.11. Results

Five runs from different initial populations were carried out at each value of span (10, 12, ... 22 m) as shown in Table 2.5. The results are reasonably consistent with the active constraints tending to be either deflection or the maximum bending moment in the longitudinal members.

Table 2.5: Results of grillage optimisation

Span	run	N_x	N_v	I_l	I_t	vol (m^3)	generation	design criterion
10	1	4	3	0.0074	0.0054	12.76	11	deflection
	2	5	3	0.0054	0.0054	13.32	13	
	3	4	3	0.0074	0.0054	12.76	9	
	4	5	3	0.0054	0.0054	13.32	4	
	5	4	3	0.0074	0.0054	12.76	2	
12	1	5	3	0.0171	0.0074	24.26	9	
	2	5	3	0.0171	0.0054	23.52	13	
	3	3	3	0.0273	0.0074	19.64	1	
	4					20.46	0	
	5	3		0.0273	0.0054	18.90	12	deflection
14	1	3	3	0.0417	0.0054	25.32	20	deflection
	2	3	3	0.0417	0.0100	26.88	1	
	3	3	3	0.0417	0.0054	25.32	20	
	4	3	3	0.0417	0.0054	25.32	7	
	5	3	3	0.0417	0.0054	25.32	13	
16	1	3	3	0.061	0.0054	33.36	3	M_x
	2	3	3	0.061	0.0074	34.10	9	
	3	3	3	0.061	0.0054	33.36	17	
	4	3	3	0.061	0.0054	33.36	14	
	5	3	3	0.061	0.0054	33.36	10	
18	1	3	5	0.086	0.0054	46.08	25	M_x
	2	3	5	0.086	0.0054	46.08	15	
	3	3	5	0.086	0.0054	46.08	21	
	4	4	5	0.086	0.0054	59.04	9	
	5	3	5	0.086	0.0054	46.08	23	
20	1	5	3	0.086	0.0054	76.32	4	deflection
	2	5	3	0.086	0.0054	76.32	7	
	3	5	3	0.086	0.0054	76.32	9	
	4	5	3	0.086	0.0054	76.32	20	
	5	5	3	0.086	0.0054	76.32	7	
22	1	8	3	0.086	0.0054	131.04	4	
	2	8	3	0,086	0.0054	131.04	10	
	3	8	3	0.086	0.0054	131.04	4	
	4	8	3	0.086	0.0054	131.04	4	
	5	8	3	0.086	0.0054	131.04	4	M_x

2.23. OBSERVATIONS AND CONCLUSIONS

We conclude the chapter with some general observations.

1. Population-based methods are demanding on computer memory and processing speed. Careful planning is essential in the setting-up of the GA if demands on memory are not to be excessive. The dimensions of the main arrays are controlled principally by population size, number of design variables and lengths of the binary strings. The attractiveness of the GA has increased with continuous improvement in computer processing capacity and capability.

2. Adaptation of the GA controls is essential with some experimentation necessary to achieve efficient operation. In large combinatorial problems some space-reduction device is necessary.

3. A sufficient number of runs must be used to obtain consistent results.

4. Structural reanalysis by trained neural network is convenient for integration into the GA allowing automatic updating of the design data.

ACKNOWLEDGEMENTS

The author expresses his appreciation of the considerable help and encouragement provided by the Department of Mechanical Engineering, Heriot-Watt University (Professor B. H. V. Topping), the Department of Civil Engineering, University of Leeds and the Department of Computer Science, University of York (Dr. J. Austin and staff).

REFERENCES

1. Goldberg D E: Genetic Algorithms in Search, Optimization and Machine Learning. Adison Wesley 1989

2. Davis L: "Handbook of Genetic Algorithms", Van Nostrand Reinhold, New York, 1991.

3. Mitchell M: "An Introduction to Genetic Algorithms", A Bradford Book, MIT Press, Cambridge, Massachusetts, 1996

4. Jenkins W M: Towards structural optimization via the genetic algorithm. Computers and Structures. 1990 40(5), 1321-1327.

5. Jenkins W M: (1991) Structural optimization with the genetic algorithm. The Structural Engineer, London, 69(24), 418-422.

6. Jenkins W M: (1992) Plane Frame Optimum Design Environment Based on the Genetic Algorithm. Journal of Structural Engineering. Proceedings of the American Society of Civil Engineers, Vol 118 No 11 November 1992.

7. Jenkins W M: (1998) Improving structural design by genetic search. Computer-aided Civil and Infrastructure Engineering 13(1998) 5-11.

8. Chakrabarty B. K: "A model for optimum design of reinforced concrete beams" ASCE   Journal of Structural Engineering, Vol 118 No 11 November 1992.

9. Jenkins W. M: Discussion on reference [8]. ASCE Journal of Structural Engineering, March 1994.

10. Levy, Stephen: "Artificial Life" Penguin Books 1993.

11. Sigmund, Karl: "Games of Life" Penguin Books 1993.

12. Hasancebi O and Erbatur F. Evaluation of crossover techniques in Genetic Algorithm based structural optimization.   Advances in Engineering Computational Technology, Ed. B H V Topping, CIVIL-COMP Press, Edinburgh 1998.

13. Jenkins W.M: "An enhanced genetic algorithm for structural design optimization", 5th International Conference on Civil & Structural Engineering, Heriot-Watt University, Edinburgh, August 1993. Civil-Comp Press, Edinburgh, 1993.

14. Davis, L: "Adapting operator probabilities in genetic algorithms". Proc. Third Int. Conf. On Genetic Algorithms, Morgan Kaufmann, Los Altos, CA, 1989.

15. Jenkins W.M: "A space condensation heuristic for combinatorial optimization", Advances in structural optimization, Computational Structures Technology 94, Civil Comp Press, Edinburgh, UK 1994.

16. Reeves C.R: (ed), "Modern Heuristic Techniques for Combinatorial Problems". McGraw-Hill, Maidenhead, Berkshire,UK, 1995

17. Schoenauer M: "Shape Representations for Evolutionary Optimization   and identification in Structural Mechanics",   Genetic Algorithms in Engineering and Computer Science, Ch 22, pp 443-463, John Wiley 1995.

18. Jenkins W.M: "On the applications of natural algorithms to structural design optimization". Engineering Structures, Vol 19, No. 4, pp 302-308, 1997 Elsevier Science Ltd.

19. Baker J.E: "Adaptive selections methods for genetic algorithms". In J.J. Grefenstette, ed. Proceedings of the First International Conference on Genetic Algorithms and Their Applications. Erlbaum.

20. Press W H, Teukolsky S A, Vettering W T, Flannery B P. Numerical Recipes in C, Second edition, Cambridge University Press 1992.

21. Krishnakumar K, Narayanaswamy S and Garg S: Solving Large Parameter Optimization Problems Using a Genetic Algorithm with Stochastic Coding. Genetic Algorithms in Engineering and Computer Science. Eds Winter G, Periaux J, Galan M and Cuesta P. John Wiley & Sons. Chichester, 1995

22. Wassermann P.D: Advanced methods in neural computing. Van Nostrand Reinhold. New York, 1993.

23. Fu LiMin: "Neural Networks in Computer Intelligence". McGraw-Hill, New York, 1994.

24. Jenkins W.M: "A Genetic Algorithm for Structural Design Optimization" in Emergent Computing Methods in Engineering Design, Eds Grierson D.E and Hajela P, NATO ASI Series F: Computer and Systems Sciences, Vol. 149 1996. Pp 30 ..53.

25. Jenkins W.M: "The Estimation of Partial String Fitnesses in the Genetic Algorithm." Developments in Neural Networks and Evolutionary Computing. CIVIL-COMP Press, Edinburgh UK, 1995, pp 137-142.

26. Koumousis V.K: "Genetic Algorithms in Optimal Design of Civil Engineering Structures". in Emergent Computing Methods in Engineering Design, Eds Grierson D.E and Hajela P, NATO ASI Series F: Computer and Systems Sciences, Vol. 149 1996. pp 54-73.

27. Osyczka A: "Multicriterion Optimization in Engineering", Ellis Horwood Series in Mechanical Engineering, Ellis Horwood Ltd, Chichester, West Sussex 1984.

28. Shih C J and Yu K C. "Methods of pairwise comparisons and fuzzy global criterion for multi-objective optimization in structural engineering". Structural Engineering and Mechanics, Vol 6, No 1 (1998) 17-30.

29. Cheng F Y and Li D. "Multiobjective Optimization Design with Pareto Genetic Algorithm". ASCE Journal of Structural Engineering, September 1997.

30. Sturzaker D. "Elementary Probability". Cambridge University Press 1994.

31. Feller W. "An introduction to probability theory and its applications". Wiley, New York 1970.

32. Whitely D: "Genetic Algorithms and Neural Networks" in Winter G A et al (eds). Genetic Algorithms in Engineering and Computer Science. John Wiley & Sons, Chichester, England 1995. pp 203 - 216

33. Jenkins W.M: "A Neural Network trained by Genetic Algorithm." International Conference of Computational Structures, Budapest, August 1996. CIVIL-COMP Press, Edinburgh.

34. Jenkins W.M: "Approximate analysis of structural grillages using a neural network." Proc. Instn Civ. Engrs Structures & Buildings, 1997, 122, Aug.,pp. 355-363

35. Braun H and Zagorski P: "ENZO-M - A Hybrid Approach for optimizing Neural Networks by Evolution and Learning." Lecture Notes in Computer Science 866. Parallel Problem Solving from Nature - PPSN III. Intl. Conf. on Evolutionary Computation, Jerusalem, Israel, October 1994, pp440 - 451.

36. Doorly D: Parallel Genetic Algorithms for Optimization in Computational Fluid Dynamics, ibid pp 251 .. 270.

37. Marshall S.J and Harrison R.F: 'Optimization and training of feed-forward neural networks by genetic algorithms.' Proc. 2nd IEE International Conference on Artificial Neural Networks, Vol 349 ch 81 pp39-43, Bournemouth, England, Nov. 1991

38. Jenkins W.M   'Neural network-based approximations for structural analysis'. Proceedings of the fourth International Conference on Artificial Intelligence in Civil & Structural Engineering. CIVIL-COMP Press, Edinburgh 1995.

39. Jenkins W.M. 'An introduction to neural computing for the Structural Engineer'. The Structural Engineer, Volume 75, Number 3, 4 February 1997.

CHAPTER 3

APPLICATIONS OF NEURAL NETWORKS IN MODELING
AND DESIGN OF STRUCTURAL SYSTEMS

P. Hajela
Rensselaer Polytechnic Institute, Troy, NY, USA

ABSTRACT

There has been considerable recent activity in exploring biological motivated computational paradigms in problems of engineering analysis and design. Such computational models are placed in a broad category of soft-computing tools that span the gap between traditional procedural methods of computation on one side, and heuristics driven inference engines (non procedural methods) on the other. Of these, methods of neural computing and evolutionary search have been extensively explored in problems of structural analysis and design. The purpose of the present chapter is two-fold. It provides an overview of those neural network architectures that are pertinent to the problem of structural analysis and design, including the back-propagation network, the counterpropagation network, the ART network, and the Hopfield network. It then provides a summary of select applications of neurocomputing in the field of structural synthesis. This summary includes the applications of neural networks in function modeling, in establishing causality in design data, in function optimization, and in diagnostics of structural systems.

3.1. INTRODUCTION

Artificial neural networks were inspired by the impressive cognitive and data processing characteristics that are typical of the brain. While the functioning of the 'biological neural network' and its intricate system of sensors is not completely understood, there is some agreement that this biological machine comprises of about 100 billion threshold logic processing elements also referred to as neurons or brain cells. The number of interconnections between these neurons is estimated to be about one quadrillion. With such a high degree of complexity, it is not surprising that the similarities between the biological and artificial systems, are, at best, superficial. Nevertheless, models describing the cognitive process that have originated in neurobiology and psychology, have been embraced by computer scientists in the development of different neural network architectures.

Explained simply, the biological neuron shown in Fig. 3.1 consists of a chamber partitioned by a semi-permeable membrane, and contains a fluid with varying concentrations of K^+ and Na^+ ions. The movement of these ions across the membrane due to the electrical impulses received by the cell results in a voltage potential, and an ultimate "firing" of the neuron. This signal is then relayed to other cells to which the firing neuron may be connected. The electrical impulses are picked up by "dendrites" and the "synapses" or connections determine the strength of the signal. The stimuli is relayed to the cell by the "axon", where it may be further strengthened or inhibited. Of the brain functions patterned by various ANN architectures, learning is perhaps the most widely studied. This information is thought to be represented in a biological neural network by a pattern of synaptic connection strengths of the various neurons and the firing characteristics of the individual neurons.

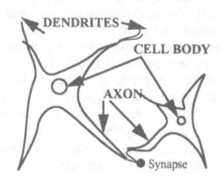

Fig. 3.1. Schematic of the biological neuron

The artificial counterpart of the biological brain is the artificial neural network (ANN), and can be described as a massively parallel, interconnected network of basic computing elements that demonstrate information processing characteristics similar to several hypothesized models of the functioning of the brain. Analogous to the biological model which consists of a large number of interconnected neurons, the ANN comprises of a number of similarly connected computational elements referred to as artificial neurons. Such models of computation are distinctly different from traditional numerical computing, and the more recently emergent strategies of symbolic processing. There has been significant recent activity in adapting this computational model in various fields of engineering. Applications have included image processing and pattern recognition, fault detection, diagnostic systems, the use of neural networks as function approximations, and formulations which allow the use of neural networks as numerical optimization algorithms. Recent interest has also centered on the use of neural networks to identify causal relations in numerical data, and this has intriguing possibilities in problems of structural design.

The chapter is organized into three distinct sections. After a brief discussion of the basic components of a neural network and its biological counterpart, those network architectures for which the training requires a supervised presentation of input and its corresponding output, are discussed, includes some details of the training algorithm, the computational effort required in network training, and the mathematical basis behind the use of these networks as tools for function approximation. The discussion then turns to networks where the training is unsupervised, with a focus on the ART and Hopfield networks. Small representative examples of applications of the ART and Hopfield network are included in this discussion. The discussion then turn to applications of the backpropagation and counterpropagation neural networks in problems of structural analysis and design.

The multilayer perceptron model is perhaps the most widely used neural network architecture. This model has been used most commonly in a mode where a trained network provides a mapping Φ between some input $X \in R^n$ and output $Y \in R^m$. Recent work has also shown that once such a network is trained, it is possible to determine the extent of influence between a component of the output vector and each input component. Details of the network architecture and the procedure necessary to train the network will be presented in this chapter. Another neural network architecture referred to as the counterpropagation network, or more specifically its improved version, has been shown to be comparably effective in function approximation. Although requiring larger amounts of training data to produce acceptable function mappings, this network is considerably easier to train than the multilayer perceptron model, and provides an additional capability of generating inverse mappings. The training algorithm used for this network will also be discussed in the chapter.

A second class of network architectures examined in this review can be broadly described as self-organizing networks; the focus of this discussion will be on the ART and Hopfield neural networks. These networks have been used in problems of vector classification, which in itself is useful in conceptual design problems. The discussion in the present chapter, however, will be on their role in direct numerical optimization. In particular, the approach has been shown to be applicable in combinatorial optimization problems, such as those resulting from the presence of discrete/integer variables in the design space. Near-optimal solutions to such NP complete problems can be generated with significantly reduced investment of computational resource.

3.2. NEURAL NETWORKS - BIOLOGICAL AND ARTIFICIAL

Artificial neural networks were inspired by the impressive cognitive and data processing characteristics that are typical of the brain. While the functioning of the 'biological neural network' and its intricate system of sensors is not completely understood, there is some agreement that this biological machine comprises of about 100 billion threshold logic processing elements also referred to as neurons or brain cells. The number of interconnections between these neurons is estimated to be about one quadrillion. With such a high degree of complexity, it is not surprising that the similarities between the biological and artificial systems, are, at best, superficial. Nevertheless, models describing the cognitive process that have originated in neurobiology and psychology, have been embraced by computer scientists in the development of different neural network architectures.

It is perhaps more productive to view the ANN technology in much simpler terms. A computer science perspective of this field is provided in a two-volume compendium [1,2], and has been referred to as Parallel Distributed Processing (PDP). ANN's are typical of a case where a biological analogy was used as a motivation for an artificial counterpart in the earlier stages of development, and once successful, the latter developed an evolutionary path of its own, departing from its biological counterpart. The artificial entity corresponding to a biological neuron is shown in Fig. 3.2. The processing element receives a set of signals X_i, i=1,2,...n, similar to the electro-chemical signal received by a neuron in the biological model. In the simplest implementation, the modeling of synaptic connections is achieved by multiplying the input signals by connection weights w_{ji} (both positive and negative). The effective input to each processing element is therefore obtained as follows:

$$Z_j = \sum_n w_{ji} X_i \qquad j = 1, J \qquad (3.1)$$

In a neurobiological system, the neuron fires or produces an output signal only if the combined input stimuli to the cell builds to a threshold value. In the artificial counterpart, this effect is simulated by processing the weighted sum of the inputs by an activation function F to obtain an output signal $Y_j = F(Z_j)$.

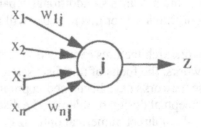

Fig. 3.2. The artificial neuron

Although various forms of activation functions have been proposed, the most commonly used sigmoid function is given by the expression

$$F(Z_j) = \frac{1}{\exp(-Z_j + T_j)} = Y_j \tag{3.2}$$

where, T_j is a bias parameter used to modulate the element output. A sketch of this function and two other activation functions are shown in Fig. 3.3. The principal advantage of this function is its ability to handle both large and small input signals. The slope of the function is representative of the available gain. For both large positive and negative values of the input signal, the gain is vanishingly small; at intermediate values of the input signal, the gain is finite. Hence, an appropriate level of gain is obtained for a wide range of input signals. The output obtained from the activation function may be treated as an input to other neurons in the network.

In principle, artificial neurons can be interconnected in any arbitrary manner. One classification is based on whether the flow of the stimuli is from the input to the output nodes only, or if neurons can also relay stimuli backwards to activate or inhibit neuron firings. Feedforward networks, characterized by the flow of stimuli in one direction only, will be discussed under the category of networks where the training is supervised. The multilayer perceptron model and the counterpropagation networks are specific architectures to be explored. The second class of networks where information is permitted to flow backward as well, are the recurrent networks, and will be discussed in the context of the Hopfield network.

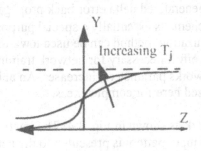

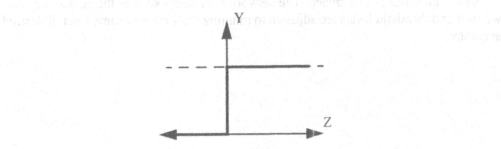

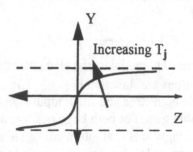

Fig. 3.3. The sigmoid, threshold and hyperbolic tangent activation functions

3.2.1. Backpropagation Neural Network

The simple feedforward networks have a layer of neurons to which the external stimuli are presented, a series of hidden layers, and a layer of neurons at which the output is available. The input neurons do not process the input stimulus; they simply serve as "fan-out" points for connections to neurons in successive layers. The presence of the hidden layer, and the nonlinear activation functions, enhance the ability of the networks to learn nonlinear relationships between the presented input and output quantities. This "learning" or "training" in feedforward nets simply requires the determination of all interconnection weights w_{ji} and bias parameters T_j in the network. Once such a trained network is established, it responds to a new input within the domain of its training by producing an estimate of the output response. Variations of the generalized delta error back propagation algorithm have been used for this training; this scheme is essentially a special purpose steepest descent algorithm, and indeed, any optimization method can be used towards this end. The only concern would be the computational effort necessary for network training when the network size (number of independent networks parameters) increases. An abbrieviated description of the training process is summarized here for completeness.

Consider a three-layer network as shown in Fig. 3.4. The interconnection weights are first initialized randomly and the input pattern is presented to the network from which the network output is determined. This output is compared to the expected output, and a sum of the squares of all errors is determined. The network parameters such as the interconnection weights and threshold levels are adjusted to minimize this error to some level of desired accuracy.

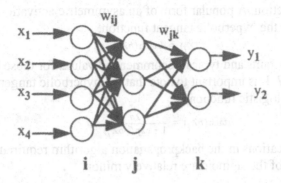

Fig. 3.4. A three-layer feedforward neural network

Let Y_j be the output at the j-th neuron of the output layer, and Y_j^T be the expected or target value of this output from the chosen training pattern. The error at training step 't' is then determined as

$$e_j(t) = Y_j^T - Y_j \tag{3.3}$$

and the sum of squared errors over all neurons in the output layer are determined as follows:

$$E(t) = \frac{1}{2} \sum_{j \in J} (e_j(t))^2 \tag{3.4}$$

In the above, J is the total number of neurons in the output layer. The backpropagation (BP) algorithm provides an incremental change to the weights w_{ji} in proportion to the instantaneous gradient Using the chain rule of derivatives, one can write the correction to the interconnection weights as follows:

$$\Delta w_{ji} = -\eta \frac{\partial E(t)}{\partial w_{ji}} \tag{3.5}$$

In the above, η is the learning rate parameter and is indicative of the rate with which the weights are changed at any given learning cycle. Note that this can be computed for any neuron for which the error $e_j(t)$ can be explicitly computed, and this is possible only for the output layer where the error can be specifically known as the difference between the network predicted and known (target) output. In order to apply this weight change algorithm to neurons in the hidden layer (for which there is no known output), the error signal must first be back-tracked to that hidden layer. For such hidden layers, the expressions for weight correction [3] is obtained as follows,

$$\Delta w_{ji} = -\eta \left(\frac{\partial F_j}{\partial Z_j(t)} \sum_{k \in J} e_k(t) \frac{\partial e_k(t)}{\partial Z_k(t)} w_{kj}(t) \right) Y_j(t) \tag{3.6}$$

and requires knowledge of the error signals for all neurons that lie to the immediate right of the hidden layer neuron 'j', and are connected to 'j'.

The training of the multilayer perceptron model using the BP algorithm can be improved by selecting an activation function which is asymmetric, as opposed to the nonsymmetric sigmoidal activation function. A popular form of an asymmetric activation function is a modified sigmoid given by the hyperbolic tangent function,

$$F(Z_j) = a\tanh(bZ_j) \tag{3.7}$$

where a and b are constants and typical recommended values for these constants are a=1.716 and b=0.6667. It is important to note that the hyperbolic tangent function is simply a biased and rescaled logistic function,

$$a\tanh(bZ_j) = \frac{2a}{1 + \exp(-bZ_j)} - a \tag{3.8}$$

and hence, the modifications in the backpropagation algorithm required in implementing this function in place of the sigmoid are relatively minor.

There are essentially two modes of training that can be implemented. The first, referred to as the pattern mode updates the weights after each training pattern is presented. So each cycle of training involves weights to be changed a total of NPAT times, where NPAT is the total number of training patterns. The order in which these training patterns are presented to the network can be changed from one training cycle to another at random, giving the process some mechanism to avoid converging to a relative optimum in the weight space. A second approach, referred to as the batch mode, computes a cumulative error computed over all of the training patterns, and makes weight changes in proportion to the rate of decrease of this cumulative error. The relative effectiveness of the two approaches is problem dependent although the pattern mode generally performs better. It is worthwhile to add that it is sometimes possible to avoid being trapped into a relative minima of the error function by adding to the weight increment a term that is proportional to the weight change in the previous cycle of training. This term is referred to as the momentum term, and the constant of proportionality used to implement this addition is defined as the momentum coefficient.

Other helpful insights into the training of such networks include the following.

- The desired response of the trained network available at the output neurons should be offset by some small amount away from the limiting values (positive or negative) of the activation functions. This prevents neuron saturation and a considerable improvement in training eficiency.
- All interconnection weights and bias constants should be set at initially small values, randomly distributed within some narrow range. This is to prevent saturation of neurons early in the training process, resulting in slow learning.
- In some cases, it is beneficial to set the learning rate of each neuron individually, the weight adjustments are in proportion to the error gradients, and these gradients can vary significantly in magnitude. While, in general, an approach that uses lower learning rates for the layers closer to the output layer, with increasing rates of learning towards the input layer has the desired effect, these learning rates can be derived using a one-dimensional search. The latter comes at some computational cost, and the user must weigh the associated benefits of this additional computation.

The reader also needs to be aware of a potential problem in network training that is referred to as overtraining. This is a phenomenon wherein the training error is forced to such low levels that the network essentially memorizes all the training patterns, with a very serious loss in generalization capabilities. A process known as cross-validation (derived from statistics) is recommended during training, wherein the available training set is split into a training and a validation set. As the training proceeds, the validation samples are used to ensure that the generalization error continues to decline with the training error. When the trend reverses, and if the network training error is within acceptable bounds, the network is considered to be trained. Typically, 10% to 20% of the training samples should be set aside for cross validation purposes. Cross validation can also be used to study the effect of varying the network architecture.

When the BP neural network is used to map input-output functions, it is representing a special form of response surface where the response function is a nested squashing function; the interconnection weights correspond to the regression parameters in a typical regression model. The polynomial regression models are good for linear mappings but difficult, if not impractical for large dimensionality, nonlinear mappings. If for example the matrix inversion approach [4] is used to solve for the regression parameters, then with NPAR parameters, the modeling capacity would be approximately proportional to the square root of the memory size. The neural network approach has no such restriction and the modeling capacity would be proportional to the actual memory size. In other words, if a polynomial model using inverse methods is used to derive a mapping, it could afford at most NPAR parameters while the BP neural network would allow the use of NPARxNPAR parameters.

In using a neural network for function response approximation, a number of associated issues must be resolved. First, a set of input-output training patterns must be obtained from the real process that one is attempting to simulate. Determination of the number of such training pairs, and how they should span the intended domain of training, requires experimentation and experience. The same statement is applicable to the selection of the network architecture, i.e., the number of hidden layers and the number of neurons in such layers. One approach to this problem is to assign a network architecture in accordance with the physical complexity of the problem, and to then determine the size and distribution of the training set. The second approach would be to fix the number of training patterns and to determine an optimal network architecture. The objective in either of these approaches is to both facilitate the network training and to ensure that the network generalizes effectively, i.e., the input-output relationship computed by the network is "correct" to within some tolerance levels. In practice, a bound on the size of the training set NPAT for good generalization is stipulated as $NPAT > \dfrac{W}{\varepsilon}$, where ε is the bound on the generalization error, and W is the number of interconnection weights in the network. So, with an error of 10%, the number of training samples should be ten times the number of interconnection weights. This is based on a distribution-free, worst case formula for a single-layer network, and as will be shown in subsequent examples, for many problems in engineering, the number of patterns required for good generalization is well below this bound. Recent developments in machine learning and computational intelligence also indicate that there may be formal methods of discovering the distribution of training patterns and/or network architecture [5].

Barron [6] established bounds on the approximating properties of a multilayer perceptron. The mean-squared error between the target function and its estimate is bounded to be of order

$$O\left(\frac{C_f^2}{M}\right) + O\left(\frac{Mp}{NPAT}\log NPAT\right) \tag{3.9}$$

where C_f defines the regularity of the target function f and is the first absolute moment of the Fourier magnitude distribution of f. Also, M is the number of hidden neurons, $NPAT$ is the number of training patterns, and p is the number of input nodes. Note that while the first term which is indicative of the best fit requires a large M, the second term which measures the empirical fit actually requires a small $\frac{M}{NPAT}$ ratio; exponentially large sample sizes are really not required in neural network based approximations.

3.2.2. A Modified Counterpropagation Network

The counterpropagation (CP) neural network was first introduced by Hecht-Nielsen[7] as a combination of two basic architectures - the Kohonen's self-organizing neural network and Grossberg's outstars neurons. This architecture required less computational effort to train than the multilayer perceptron architecture described in the previous section. Training times are of considerable importance when one considers modeling of extremely large structural systems. However, the original version of the network did not receive widespread attention due to its unimpressive generalization performance, particularly in comparison to the multilayer perceptron model. As shown in Fig. 3.5, this network contains three layers - a fan-out layer as in the BP network, a layer of Kohonen or feature sensitive neurons, and an interpolating or Grossberg layer. The inputs to the network are directed to the Kohonen layer, which acts like a clustering device. In other words, neurons in this layer classify all input vectors based on some identifiable features in these vectors. Each neuron in the Kohonen layer represents one such cluster, and the interconnection

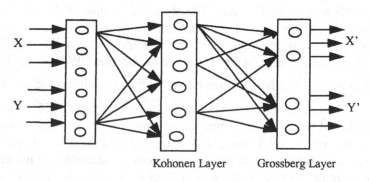

Fig. 3.5. Schematic of a counterpropagation neural network

weights between the input nodes and this neuron are representative of an average of all input patterns of that cluster. Similarly, the interconnection weights between each Kohonen neuron and the output or Grossberg layer neurons are representative of an averaged output of all patterns belonging to the cluster. If the radius of each cluster is infinitesimally small, then each Kohonen neuron will only represent one pattern, and the network would be a memory record of all input-output patterns. On the other hand, if the radius of the cluster is large, then several patterns would be classified in the same cluster. A larger radius generally results in significant errors during generalization, as an input pattern presented to the network is classified as belonging to a cluster on the basis of its similarity to the stored cluster weights for each Kohonen neuron. Even if the input pattern were classified into the right cluster, the output would only be an average of the outputs of all input patterns belonging to that cluster, and that were used in network training.

The unsupervised training of the Kohonen neuron is based on a minimum disturbance principle in which the weight vector of only one neuron, the one closest to the current input, is modified. When a new input vector activates one of the Kohonen neurons, the weight vector representing the connection between this neuron and the outstars is modified (in the training mode) or simply returned as an output (in the generalization mode). In the modified CP network described here, the output is actually a nonlinear blend of outputs from several Kohonen neurons which a given input may activate, albeit to different degrees. This nonlinear averaging allows for better generalization characteristics from the network.

As for the BP network, this network may be used to develop the mapping,

$$\phi|((X \in R^n) \to (Y \in R^m)) \tag{3.10}$$

and requires that weights of two layers be determined. As a consequence, there are two sets of weights to be considered - a set of Kohonen layer weights for the i-th Kohonen neuron,

$$w_i \equiv [w_{1i}, w_{2i}, \cdots w_{ni}] \in R^n, \; i = 1, k \tag{3.11}$$

and

$$z_i \equiv [z_{i1}, z_{i2}, \cdots z_{im}] \in R^m, \; i = 1, k \tag{3.12}$$

for the Grossberg layer. The training algorithm for this network is discussed in [8]. It suffices for the purpose of this discussion to recognize that the training algorithm increases the size of the network, i.e., the number of Kohonen neurons k, if the new input pattern is more than δ_r away from the already defined neurons, or, it modifies the i-th neurons of both layers which are the closest in the absolute value norm sense to the currently presented input pattern.

Since a look-up table form of the network would result in a very large number of Kohonen neurons to meet requirements of good generalization, an output interpolation is performed.

Consider some input vector, $\hat{x} \in R^n$ is presented to the network. The best match for this vector among all k weight vectors is found, and the distance between this vector and closest Kohonen neuron is computed as δ_o. Next, an index set S is determined that contains all indices s for which the following relation holds:

$$\delta_s = \sum_{l=1}^{n} |\hat{x}_1 - w_{ls}| \le \delta_o + \delta \tag{3.13}$$

Here, the arbitrary parameter δ defines a neighborhood of the winning neuron, and is defined as the network interpolation parameter. Assuming that τ such neurons are found to be in the neighborhood, the output of each is used to develop the network output. To compute an individual contribution, h_s of each such neuron to the output, a membership function $f(d_s)$ is used. This function is defined as a power function

$$f(d_s) = 1 - d_s^r \tag{3.14}$$

where d_s is a normalized distance calculated as follows:

$$d_s = |\delta_o - \delta_s| / \delta \tag{3.15}$$

The contributions h_s of all τ neurons are normalized as

$$h_s = \frac{1 - d_s^r}{\sum_s 1 - d_s^r} \tag{3.16}$$

so that they add up to unity, and the output is simply computed as

$$out = \sum_{s=1}^{\tau} h_s z_s \tag{3.17}$$

Such network response, developed through interpolation, tends to yield better results as the sharp boundaries between clusters are removed. Clearly, the number of contributing neurons, τ depends on the value of δ, set during testing to minimize the approximation error, and may vary for different input vectors. In a case where the input neuron is very close to an input vector, it becomes the sole contributor to the output response. The shape of the membership function is controlled by the exponent r, and its value must be adjusted for the problem at hand so that the approximation error is minimized.

Examples pertaining to the use of the CP networks in a variety of different applications in structural design, will be presented in a later section. We next examine another network architecture, referred to as the Hopfield model, which belongs to the general category of recurrent networks.

3.2.3. The Hopfield Network

Networks that have a feedback path between the output and input neurons are described as recurrent networks, and are said to provide a more realistic modeling of the memory process. Such networks also exhibit a form of learning referred to as self-organization, such as may be required if no known output is available to train the network. The Hopfield network is a widely studied network architecture where the self-organizing behavior is exhibited. A schematic sketch of this network architecture is shown in Fig. 3.6. The network architecture is seen to contain the two principal forms of parallel organization that are characteristic of neural systems, viz., parallel treatment of inputs and outputs, and extensive connectivity between the neurons. As can be seen from the figure, the i-th neuron receives an input which

is a sum of externally applied inputs I_i, and a weighted sum of outputs V_j from the other
neurons.

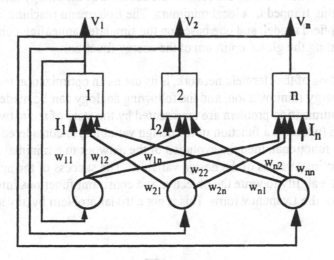

Fig. 3.6. The Hopfield Network

The state of the Hopfield network is the value of the outputs of its neurons. For the binary
Hopfield network, each neuron can have a value of 0 or 1, and for a network with 'n' neurons, there are 2^n states for the network, each characterized by a set of weights and bias
inputs. In network training, these connection weights and bias inputs must be established.
The state of the network can be used to define an energy function as follows:

$$E = -\frac{1}{2}\sum_i\sum_j w_{ij}V_iV_j - \sum_i V_iI_i \tag{3.18}$$

Here, the output of the j-th neuron denoted by V_j is obtained by processing the sum of all
inputs to that neuron, u_j, through a gain function g_j as follows:

$$V_j = g_j(u_j) \tag{3.19}$$

The change in the energy of the network due to a change in the state of neuron j can be written in the following form:

$$\delta E = -\left(\sum_i w_{ij}V_i + I_j\right)\delta V_j \tag{3.20}$$

The system is considered to have reached a locally stable configuration when its energy
does not change. When Hopfield networks are to be used as an associative memory device,
the interconnection weights must first be obtained during the training process by presenting
known inputs to the system. Once these weights are determined, they are frozen and used
for subsequent classification of other inputs. When a new input vector is presented for classification, the various stored states are searched to determine the one closest to the input.
The dynamics of this search are derived on the basis of determining the minimum of an
energy functional, where the latter is a function of the network state, the interconnection
weights, and the bias input. For stable network dynamics, the energy functional must be of

the Lyapunov form. The input pattern to be classified is presented to the network, and the latter is allowed to relax along the energy contours, settling to a state that is most similar to the presented input. Since the network follows the contours of the energy function, there is a possibility of getting trapped in a local minimum. The Boltzmann machine, a stochastic variant of the Hopfield model, and one based on the simulated annealing concept, has a better chance of locating the global optimum of the energy function.

A second application of the Hopfield network, is its use as an optimization tool. The process is again one of energy minimization, and the following analogy can be made. If the design variables of an optimization problem are represented by the state of the network, and the objective function (which is a function of the design variables) is considered to be equivalent to the energy functional, then the evolution of the network to a minimal energy state would yield the optimal values of the design variables. The success of the approach is dependent upon the ability to write the objective and constraint functions into an energy functional that is of the Lyapunov form. This is not a trivial problem by any stretch of imagination.

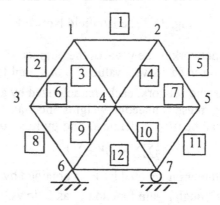

Fig. 3.7. A 12 bar planar truss structure

Consider the following application of Hopfield networks in an optimal design problem involving a truss assembly (Fig. 3.7) that may be assumed to consist of a few groups of equal length members. In each group of elements and joints, there may exist errors in lengths and joint diameters, where such errors are typically small in relation to nominal values of these dimensions. Further, it is assumed that the member lengths and joint diameters for a given truss can be determined precisely, and that errors in joint diameters can be represented as equivalent member length errors.

This problem is essentially one of optimal assignment, where a particular member is assigned to a special position so as to reduce the overall shape distortion and to minimize the member pre-loads. To map this problem into the energy space corresponding to a Hopfield network, a variable v_{xi} is defined, the value of which is unity when member x is assigned to position i, and zero otherwise. If each neuron (binary) in the Hopfield network is assigned to one design variable, the state of the network describes the design variable set at any stage of the network evolution. As shown in [9], the energy functional of the Hopfield network can be expressed in terms of the neuron outputs and the interconnection weights in the desired Lyapunov form. This energy functional contained four terms, one related to the total distortion produced from errors in element lengths, and three penalty terms associated with violation of constraints that each member is uniquely assigned to one position in the truss structure. In terms of the variable V_{xi}, a mathematical statement of the optimal placement problem may be obtained as follows:

$$
Minimize \ (E) = \begin{pmatrix} \dfrac{A}{2}\sum_x\sum_i\sum_{j\neq i} V_{xi}V_{xj} + \dfrac{B}{2}\sum_i\sum_x\sum_{y\neq x} V_{xi}V_{yi} + \\[2mm] \dfrac{C}{2}\sum_x\left(\sum_i V_{xi} - 1\right)^2 + \dfrac{C}{2}\sum_i\left(\sum_x V_{xi} - 1\right)^2 + \\[2mm] \dfrac{D}{2}\sum_{r=1}^{NDOF}(U_r)^2 \end{pmatrix} \tag{3.21}
$$

where U_r is the displacement in the r-th dof, and $NDOF$ are the total number of degrees of freedom in the structure. In the above, the first four terms are required to account for the constraints that stipulate that only one member be assigned to each position, and that all 'n' available positions are assigned an element. The last term simply represents the objective function representing a minimal distortion of the truss structure. Note that V_{xi} can be zero or unity, indicating the absence/presence of a member x at location i. The network weights and the state of the system must evolve so as to generate the minimum energy state. This would correspond to an optimal assignment of the available members. The evolution of the neurons is defined by the following equation,

$$
u_{xi}^{(k+1)} = u_{xi}^{(k)} + \Delta u_{xi} \tag{3.22}
$$

where,

$$
\Delta u_{xi} = \frac{\partial E}{\partial V_{xi}} -
$$

$$
A\sum_{j\neq i} V_{xj} - B\sum_{y\neq x} V_{yj} -
$$

$$
C\left(\sum_j V_{xj} - 1\right) - C\left(\sum_y V_{yi} - 1\right) - \tag{3.23}
$$

$$
D\sum_{r=1}^{NDOF} U_r \frac{\partial U_r}{\partial V_{xi}}
$$

Comparing eqn. (31) with an expression for the input to neuron xi given as,

$$
\Delta u_{xi} = \sum_y\sum_j w_{xi,\,yj} V_{yj} + I_{xi} \tag{3.24}
$$

one can derive expressions for the updated weights $w_{xi,\,yj}^{(k)}$ and the input bias terms I_{xi}. These weights are then used with the old V_{yj}'s and I_{xi}'s to determine the new u_{xi}, which is then processed through the gain function to obtain the new V_{xi} for a second round of network evolution. The process is stopped when the V_{xi} are close to their admissible values of 0 or 1. For a stable network, successive iterations produce smaller and smaller changes in the output, with the system settling to a stable state. Note that in this problem, the space over which the energy function is minimized in the limit may be interpreted geometrically as the 2^{N^2} corners of the N-dimensional hypercube defined by the binary values of V_{xi}.

This approach has also been used as an optimization tool to perform node numbering in a finite element analysis to minimize the bandwidth of the resulting stiffness matrix. The energy function was defined in terms of a variable V_{ij}, the value of which was one when node i was in the j-th place in the numbering scheme, and zero otherwise. It included a weighted sum of the constraints and the objective of requiring a minimal bandwidth. Good results were reported for relatively small problems; however, even for these problems, the network behavior was shown to be extremely sensitive to the choice of weighting constants used in the definition of the energy functional.

3.2.4. The ART Network

The representation of human memory, albeit in a limited form, by an artificial neural network, has been the subject of extensive research [10-12]. A special feature that must be incorporated in this model of memory is that it must allow for recognition of previously encountered patterns, and at the same time accommodate and store new input without destroying the existing contents. This stability-plasticity requirement has contributed to the proposition of adaptive resonance theory, or ART, networks.

ART networks also belong to the general category of self-organizing or unsupervised-learning systems. ART networks are essentially distinguished on the basis of the form of input information they can accept (binary or continuous), and on the approach used to process this information. The essence of the approach can be explained most simply by considering a network that can only process binary input patterns, and discussions here will be limited to this model. As stated earlier, ART networks attempt to model the process by which humans learn and recognize invariant properties of a given problem domain. A special characteristic of such networks is the plasticity that allows the system to learn new concepts, and at the same time retain a stability that prevents destruction of previously learned information. ART networks accommodate these requirements through interactions between different subsystems, designed to process previously encountered and unfamiliar events, respectively. Before discussing the application of these networks in structural design, a brief description of the network will be presented.

A schematic representation of an ART network architecture is shown in Fig. 3.8. As stated earlier, these networks function as vector classifiers, determining if an input vector presented to the network has features that are similar to those of one of the stored patterns.

There are essentially five interacting subsystems that are required to model the vector classification process.

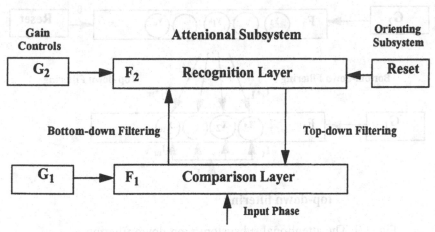

Fig. 3.8. Schematic of the ART network

As shown in the figure, these interacting subsystems are a set of two gain controls, an attentional subsystem consisting of two layers of neurons referred to as the comparison and the recognition layers, respectively, a vigilance control, and a reset control, also termed as the orienting subsystem. The attentional subsystem is designed to process familiar events. By itself, however, it is unable to process new patterns and at the same time retain stability of existing patterns. In an isolated mode, it is either rigid to new patterns or exhibits extreme instability with regard to stored patterns. This shortcoming is addressed by the orienting subsystem or the reset control. The function of each component subsystem is summarized in the following sections.

The first component subsystem is the attentional subsystem, and consists of two layers of artificial neurons - the comparison and the recognition layer. In the present discussion, these layers will be referred to as F1 and F2 , respectively. The input pattern is first processed by the F1 layer neurons, as shown in Fig. 3.9, and the output presented to the F2 layer neurons through a filter of stored weights as shown in Fig. 3.10. The F2 layer neurons are subjected to a lateral inhibition mechanism, in which the output of each neuron is fed back to itself with a positive weight (enforcement), and is presented to all other neurons in the layer with a negative weight (inhibition). This allows only one neuron in the F2 layer to fire - a "winner-take-all" strategy. The output of this activated neuron then generates a template pattern for comparison with the input pattern. A vigilance or tolerance level is assigned to this comparison, and if the two patterns are similar, the input pattern is automatically classified under the same category as the flag of the fired F2 neuron. Otherwise, a reset wave is generated from the orienting subsystem.

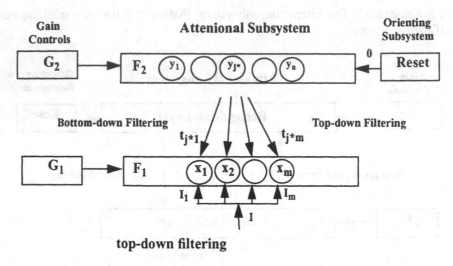

Fig. 3.9. The attentional subsystem - top-down filtering

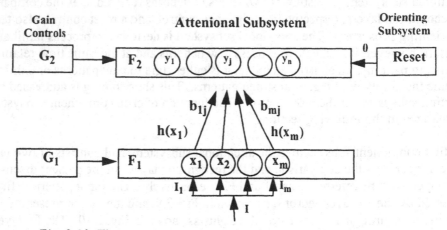

Fig. 3.10. The attentional subsystem - bottom-up filtering

The role of the orienting subsystem is primarily one of processing unfamiliar patterns. If the pattern generated by the fired F2 layer neuron is different from the input pattern, this subsystem generates a reset signal, whereby the fired neuron is disabled for a second round of presentation of the input patterns. The next dominant neuron in the F2 layer fires as a result, and the process of similarity assessment is repeated until a match is located. If no similar category can be found, as would be true for an unfamiliar pattern, a new category is established.

The gain controls Gain 1 and Gain 2 are vital for the process of training and classification. They play a central role in regulating the firing of F1 layer neurons. In the binary system described here, an F1 layer neuron fires if it satisfies the "two-thirds rule" proposed by Carpenter and Grossberg [12]. This rule states that an F1 layer neuron will fire only if at least two of the three input signals that it receives are equal to unity.

The process of vector classification is initiated with the presentation of the vector to the comparison layer. As shown in Fig. 3.9, each neuron in the F1 layer receives three binary inputs - from the input, from Gain 1, and a feedback from the F2 layer. The process is initiated by setting Gain 1 to unity and Gain 2 to zero. As a result of the latter, the feedback from the recognition layer is exactly zero. Further, since Gain 1 is unity, the two-thirds rule dictates that at the first presentation, the output of the F1 neurons will be exactly the same as the input. Associated with the j-th neuron in the F2 layer, are a set of weights b_{ij}, i=1,n and j=1,m, where n is the dimensionality of the input vector, and m is the total number of neurons in the F2 layer. As shown in Fig. 3.10, each F2 neuron receives as an input, the dot product of the vector output from the comparison layer and the weights b_{ij}. The neuron with the largest input is declared as the winner with an output unity, while outputs of all other F2 neurons is set to zero. Now that the recognition layer neurons are active, Gain 1 is set to zero. The output from the winning F2 neuron is multiplied by a set of stored binary weights t_{ji} (zero or one) as shown in Fig. 3.9, and passed back to each of the n neurons in the F1 layer. The two-thirds rule is invoked once again, and the F1 layer neurons with at least two unity inputs would fire creating a new output signal. Clearly, if the weights t_{ji} are similar to the input pattern, a match is obtained, and the input vector is classified under the current dominant neuron in the F2 layer. If there is significant dissimilarity, the previously winning F2 neuron is disabled and the process is repeated. If no match is found, a new F2 neuron weights are trained to recognize the new input pattern.

The training process ensures that the t_{ji} weights are binary valued, and that a scaled version of these weights is available as the bottom-up weights b_{ij}. These weights are representative of the different training patterns that the system has learnt to recognize. This learning is unsupervised in that there are no target patterns to emulate. Rather, the weights are developed as a result of successive reinforcement with similar patterns. Details of the network training are beyond the scope of this discussion. For the present development, it is sufficient to know that the dynamics of network learning are based on models found in psychology. This subject is discussed in greater detail in Reference 13. It is also important to indicate that ART networks are indeed amenable to parallel processor implementation. Since the stored patterns must be compared to the input pattern in sequence, the bottom-up filtering which is required can be performed simultaneously on a large number of processors. With minor modifications in the network architecture, this parallel processing would also allow for determining more than one stored pattern that possesses similar features to the input pattern.

3.3 APPLICATIONS OF ANN'S IN STRUCTURAL SYNTHESIS

We next turn our attention to some applications of the aforementioned neural network architectures in problems of structural analysis and design. These applications include generating function approximations for reduced computation time during design optimization, and, in the case of the Hopfield network, as a tool for combinatorial optimization problems.

3.3.1. ART Applications in Structural Design

Applications of the binary ART network in the conceptual design of structural systems have been considered in Reference 13. There are two distinct design processes that were considered in the present work. The first encompasses a class of structures where the structural layout was generally known and the load and support conditions were allowed to vary. The second class of problems is one where the loads and support points were assumed to be given, and the object of the design was to generate a near optimal structural topology. In either case, significant design experience with similar problems was assumed to be available, and dependent upon the identifiable critical features of the problem, a distinct procedural design process associated with the problem could be formulated. The problem may therefore be best understood as a use of ART networks to provide a memory capacity or a knowledge base for design, from which information can be recovered upon presentation of relevant features. The approach, therefore, draws upon the notion that human memory operates according to associative principles.

As a simple illustration of this idea, consider a problem in which the optimal cross section of a beam is to be determined for an allowable strength and for minimum weight. For simplicity, it may be assumed that the beam has a rectangular cross section, and further, either its cross-sectional depth or width are assumed fixed. In such a situation, the optimal cross section along the span of the beam is determined as follows:

$$h = \sqrt{\frac{6M}{b\sigma_{al}}} \quad \quad if \ b \ is \ constant$$

$$b = \sqrt{\frac{6M}{h^2\sigma_{al}}} \quad \quad if \ h \ is \ constant$$

(3.25)

This determination of optimal section properties is based entirely on the value of the bending moment at that section, where the latter is determined by the end supports and the type of loading that is applied. The type of loads and supports can thus be considered as critical features of the problem. Combination of these features result in a specific moment distribution, and hence a distinct sizing of the beam section for uniform strength. As a simple illustration, the problem features can be represented by a 4-digit binary string; the relationship between the binary numbers and the physical features is summarized in Table 3.1. A total of 16 distinct combinations can be obtained in this representation. A representative sample of these combinations and their associated procedural processes are shown in Fig. 3.11.

Table 3.1. Binary representation of problem features

A	0	The left end is a simple support
	1	The left end is a clamped support
B	0	The right end is a simple support
	1	The right end is a clamped support
C	0	The load is a force
	1	The load is a moment
D	0	The load is concentrated
	1	The load is uniformly distributed

For the comparison layer in the attentional subsystem, this problem requires a total of four nodes (one node per digit of the input pattern). Note that this discussion is confined to binary coded ART networks only. The recognition layer in this subsystem would typically have as many nodes as the number of distinct patterns presented for categorization. Reference 13 describes the results on numerical experiments with this simple problem, designed to study the influence of the vigilance parameter on the network's ability to learn and to classify the given patterns. As is to be expected, higher values of the vigilance parameter (0.65-0.75) result in the recognition of minor differences in the input pattern. This finer distinction is of increased significance as the number of problem features is increased. In the present example, the beam was divided into a number of smaller segments, and the presence or absence of a load on that segment was described by a binary variable. This subdivision may be necessary as the problem definition becomes more elaborate. Another level of refinement would be to include the magnitude of loads and moments as part of the problem features. The ART network, therefore, functions as an efficient feature classifier. Once classification of a problem is complete, a predefined procedural process takes over. New studies should be directed at looking at a continuous variable implementation of ART, and combining it with the multilayer perceptron network to include procedural processing in the distributed computing environment.

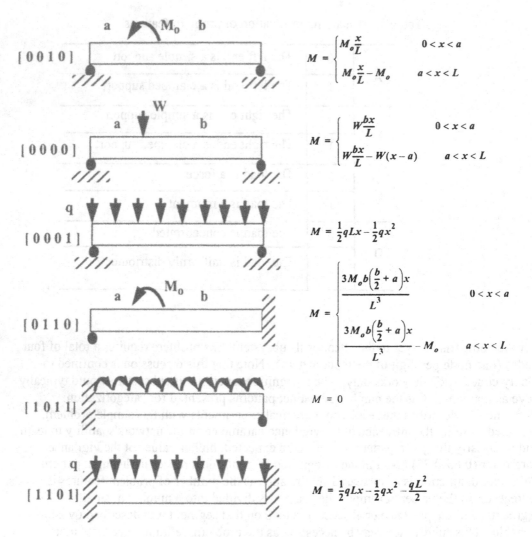

Fig. 3.11. ART codes and related procedural processes

3.3.2 BP and CP Network Applications in Structural Synthesis

This section of the chapter discusses the use of BP and CP neural networks in problems of structural analysis and design. These applications have included, among others, the use of the networks to model nonlinear structural processes, as structural engineering diagnostic systems, as a rapid reanalysis capability in optimal design, in multidisciplinary aircraft structural design, in the design of controlled structures, and as an optimal design estimation capability. A representative set of example problems are presented here to give a sense of what is possible, and what are the underlying limitations in the use of this modeling strategy. Each of these applications have drawn upon the property of such networks to model a functional relationship between some input and output quantities, given a limited set of exemplars.

3.3.2.1 BP and CP Networks in Function Approximations

The issue of function approximations has received considerable attention in the recent years. To a large extent, this interest has been fostered by increased activity in using formal optimization methods in design. Our interest in this discussion is focussed on the subject of structural optimization. This field has witnessed a healthy growth over the past four decades, and in more recent times, there has been increased interest on treating structural design within a multidisciplinary context [14-16]. This attitudinal change has grown out of a recognition that the design and development of a complex structural system can no longer be conducted by handling its different components (e.g active controls or mechanisms for loads generation) in isolation. The treatment of multidisciplinary issues in structural design is not without cost. The computational cost of the numerical solution is affected both by the inclusion of analysis from many different disciplines, and by an increase in the dimensionality of the optimization problem. The latter has a direct bearing on the number of design evaluations that must be performed in a search for the optimal solution. This situation is exacerbated by the presence of discreteness in the design space, where a search for the optimal solution is often based on stochastic search methods such as the genetic algorithm or the simulated annealing; these algorithms, while promising a greater likelihood of locating the global optimum, do require a significantly larger number of design evaluations to be performed. The logical solution to this dilemma is the use of approximate function analysis in lieu of exact analysis.

The issue of using high quality approximations in structural design is not new. In fact, Taylor series approximations have been used effectively in structural synthesis for over 25 years. As the focus of structural design has moved into a multidisciplinary realm, there has been an increased focus on examining problems where the design space may be nonconvex or disjointed, and where gradient information required to construct Taylor series approximations may be difficult to compute or simply unavailable. Furthermore, there has been a move to take multidisciplinary structural design into a realm of disciplines such as manufactureability, ease of assembly, and maintainability. Such disciplines have poorly defined or non-existent analytical models from which to compute the response required by the optimization routine. In many cases, an experiential database or observations of experimental simulations must be used to construct such models. Since gradients of such observations are both expensive and subject to errors originating in measurement and other noise, the approach is to construct approximate models based on zero-order information. In this context, the use of response surfaces has been explored extensively in problems of structural and multidisciplinary design [17-19]. As stated earlier, neural network based function mapping may be regarded as a special case of a response surface. The use of these networks to model difficult structural optimization problems is illustrated here through the use of some example problems.

The principal advantage of a trained neural network in an optimization framework over the original computational process is, that upon presentation of the input, the output can be generated with orders of magnitude less computational effort. Consequently, benefits can be substantial in those problem areas that are computationally very intensive. Note, however, that the cost of generating the training data itself must not be underestimated. Both function and sensitivity information required by an optimization algorithm can be obtained from the neural network. Fig. 3.12 describes the fundamental idea wherein a trained network is used to assist in a traditional nonlinear programming based optimal search. Following a definition of the optimization problem, the selected design variables are randomly varied over their expected range of variation, and the corresponding values of objective and constraint functions are obtained. These then comprise the input-output training pairs, and with which the network is trained.

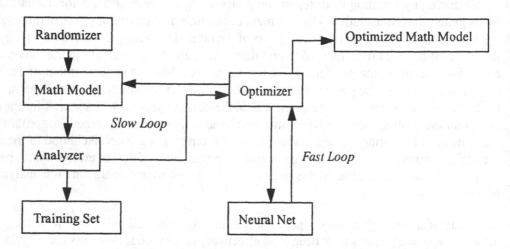

Fig. 3.12. Neural network approximations in an optimization framework

Consider a ten-bar truss of Fig. 3.13 that is to be sized for minimal weight, with constraints on y-direction displacements at nodes 3 and 5; the cross-sectional areas of the bar elements are the design variables in the problem. Presence of ten bars requires an equal number of input nodes in the network. Similarly, since the two displacement constraints are the only output quantities of interest (weight is a linear function of cross-sectional areas), the network must have four output nodes. As stated earlier, decision on the number of hidden layers and the number of neurons in each, is still something of an art, and is generally determined on the basis of past experience. Once the architecture definition is complete, the input-output data is presented to the network in the manner described earlier, and a trained network is established. This trained network can then be used as an approximate analysis tool in lieu of exact finite element based analysis. A set of results for the ten bar truss is presented in Table 3.2. For this case, the network consisted of two hidden layers with 6 neurons each (see Fig. 3.14) were used. Numerous other example problems of both lower and higher dimensionality have been studied [20,21].

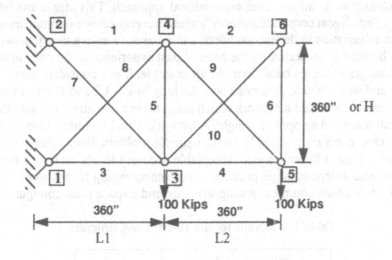

Fig. 3.13. A 10 bar planar truss structure

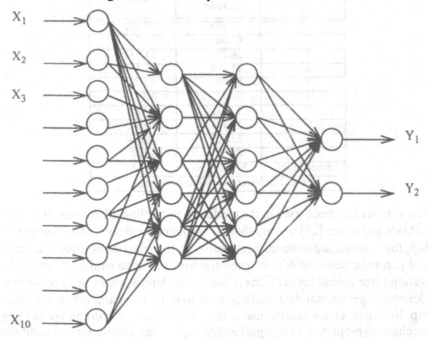

All connections are not shown for clarity

Fig. 3.14. The neural network used to model the input-output data

Experiments have also been performed to explore the idea of training a neural net to estimate optimal designs directly for a given set of design conditions, and to bypass all the analyses and optimization iterations of the conventional approach. This idea draws on the
concept of an "intelligent corporate memory", where several known optimal designs within
some domain of variation in design conditions can be used to train a neural network. This
network can be used to generalize on the basis of past experience to estimate optimal
designs for changes in design conditions. As an example, if one considers the ten bar truss
of Fig. 3.13, and would like to determine how the lengths L1, L2 and H affect the optimal
design, it is possible to create a network which maps these three quantities into the ten
cross-sectional areas and an optimal weight. A network with 14 neurons in the hidden layer
was used to achieve this mapping [22]. For this simple problem, it was shown that 10 to 15
training patterns were sufficient to yield acceptable accuracy levels for such a mapping.
This reference also describes larger problems where encouraging results were obtained,
including an intermediate complexity wing structure and a space truss configuration.

Table 3.2. Results for the 10-bar truss structure

Design Variable X_i (in^2) i	Neural Network Based Solution	Solution from Exact Analysis
1	30.508	30.688
2	0.100	0.100
3	26.277	23.952
4	11.415	15.461
5	0.100	0.100
6	0.413	0.552
7	5.593	8.421
8	21.434	20.605
9	22.623	20.554
10	0.100	0.100
Objective-lb	5010.22	5063.81
10-6-6-2 architecture, 100 training sets		

The BP network has also been used in the modeling of nonlinear behavior of structural
materials. Alam and Berke [23] discuss the use of this network as a nonlinear stress-strain
relationship, for representing material nonlinearity in the analysis of truss structures. The
modeling of nonlinear behavior of reinforced concrete has been similarly explored [24].
The motivating force behind these efforts is that for materials where the stress-strain behavior is not known, experimental data could be used to obtain a neural net based functional
relationship. Brown et. al. discuss the use of this network architecture to predict hygrother-
mal and mechanical properties of composites [25], given the environmental conditions, constituent material properties, and volume fractions of constituent materials.

The CP network has been used in a number of applications in structural engineering, including as a rapid reanalysis tool in optimal design [26], and as an inverse mapping capability in system identification problems [27,28]. One example of this latter class of problems was the use of the CP network in structural damage assessment. The static displacement response under applied loading was used to assess the extent and location of damage in the structural components. A model of structural damage proposed by Soeiro [29], based on stiffness reduction in structural components of truss and frame assemblies, was used in this study. In a typical example in this study, for a three-bay frame with 9 beam elements (see Fig. 3.15), 2×10^6 discrete damage states were considered. Of these possible states, 3600 randomly generated damage patterns were used to obtain the displacement response under the applied loading. For a prescribed network resolution parameter, a network with 504 Kohonen neurons was established, and was tested for 400 patterns of displacement that were not part of the network training. With a complete set of displacement measurements, damage estimates were established with errors ranging from 6%-10%. Even with incomplete measurements, as would be the case in real on-line damage detection where all nodal displacements cannot be measured, the pattern completion characteristics of the network allowed acceptable damage diagnostic performance.

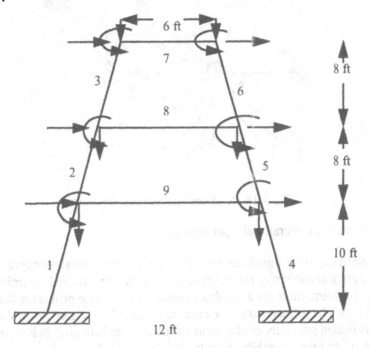

Fig. 3.15. The 3-bay frame structure

The use of the CP network as a rapid reanalysis tool [26] was tested for a particularly computationally intensive problem in structural optimization. Space truss structures with discrete variations in design variables, were sized for minimal weight and constraints on allowable stress levels in the members. Furthermore, a simulated-annealing based optimization approach was used for weight minimization; this selection was motivated by the ease with which discrete variables could be included in the optimal search. Of the 28 bars in the space truss shown in Fig. 3.16, only 15 were allowed to vary from minimum gage, and could take on any one of nine discrete values. This resulted in a design space with approximately 2×10^{14} possible design points; of these, 5000 random patterns were selected and used in the network training. A network with 1516 Kohonen neurons was established and used in the preliminary optimal search. The design space was then biased towards the preliminary design, and 5000 new patterns used to generate a second network with only about 180 Kohonen neurons. The optimization based on function estimates from the CP network required less than 10% of the CPU time that was necessary for optimal design with exact analysis.

More in-depth descriptions of successful applications of BP and CP networks in multidisciplinary structural design are presented next.

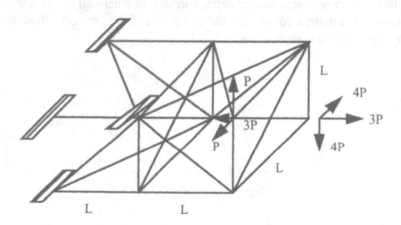

Fig. 3.16. A 28 bar space truss structure

3.3.2.2 Crashworthy Rotorcraft Subfloor Structures

A design problem requiring significant investment of computational resource deals with the subject of design of crashworthy rotorcraft structures. In using formal optimization techniques for this problem, there are a number of clearly identifiable problems that can be studied in some detail. The primary structure must be designed for a number of distinct flight loads, and this design generally conforms to elastic load deformation behavior. The structure must also be designed to exhibit a controlled collapse behavior during a crash, and provide optimal energy absorption characteristics. It has been shown [30] that the design of the structural topology in addition to structural member resizing is the more appropriate procedure to adopt. As an example, the number/placement of longitudinal beams and bulkheads in the fuselage, and their stiffness properties must both be considered as design variables to fully exploit the extent of the feasible design space. Special energy absorbing (EA) structural elements can also be designed and placed at optimal locations in the structure so as to promote a controlled collapse and distribute the energy associated with the crash in designated structural zones.

The topology design problem has been shown to result in nonconvexities in the design space, thereby reducing the effectiveness of traditional gradient-based methods for optimal design. When used, such methods have a propensity to locate an optimal design closest to the starting point. The use of global search methods such as simulated annealing or genetic algorithms offers an increased probability of locating a global optimum. However, there is a significant increase in the computational effort that is required in using these methods, and recourse to function approximation concepts becomes an important issue. Given the nature of these search strategies, local function approximations such as the Taylor series first-order approximation are of limited use. Furthermore, these methods require the availability of gradient information which, in a number of problems, is either expensive to compute, unreliable, or is simply unavailable as in discontinuous design spaces. The use of a neural network based response surface was considered to be an ideal candidate for study in this problem.

A two-stage solution strategy was adopted for the problem. In the first stage, an optimal topology and dimensions of the primary structure, the optimal placement of EA components in the subfloor, and the optimal load-deflection characteristics of these components, were established. This combined sizing-topology optimization problem was solved through the use of a neural network based global function approximation of the response function combined with a genetic algorithm based search procedure. In the second stage, for a specific EA component, the shape and dimensions were determined in accordance with the optimal load-deflection characteristics of the previous stage. For this inverse problem solution, nonlinear analysis of the EA component is done in ABAQUS [31] for a representative set of designs distributed over the design domain, and this data then used to develop a neural network based response surface that was used in the optimization procedure.

A hybrid simulation program KRASH [32] which models the major components of the vehicle as a combination of mass and nonlinear spring elements was used in simulating the crash behavior. The hybrid approach resulting from a combination of numerical and experimental techniques has been found to be invaluable in studying crash behavior for design modifications.

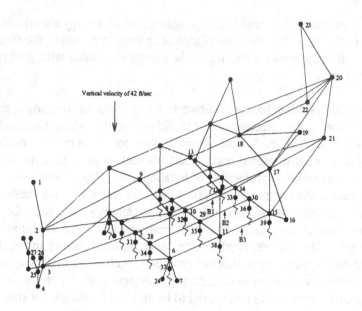

Fig. 3.17. KRASH model of a CH-47

Consider an idealized rotorcraft structure where the floor is modeled by lumped masses and linear beams, as are the pilot/co-pilot and their seats; the crushable structure is modeled by a combination of lumped masses and nonlinear spring and beam elements (Fig. 3.17). A total of 39 lumped masses and 68 linear beam elements are required for the floor, seat, and occupant model. The subfloor required 25 lumped masses, and 25 nonlinear beam and spring elements. Note that the cockpit livable space, and other major mass items and their support/ containment structure can also be modeled in a similar manner. The crash conditions may be specified as prescribed velocities, and by pitch and/or roll attitude at time of impact. In this study, the floor grillage structure formed by the intersection of longitudinal beams and bulkheads was assumed to be extremely rigid. Consequently, the points of intersection of the grillage do not exhibit significant elastic deformations, and are candidate locations for the nonlinear spring and beam elements that denote the energy absorbing crush structure. The load-deflection characteristics of the nonlinear beam and spring elements that make up the subfloor had to be provided as user input to the KRASH program.

In the stage-1 optimization problem, the stiffness distribution of the subfloor region was the design objective, and the load-deflection behavior of the nonlinear beam elements was recovered as part of the optimal design; the load-deflection data for the spring elements was kept fixed at some nominal level. The depth of the subfloor region under each energy absorbing element was a user specified input, and related to the available displacement for the crushable structure. In order to simplify the formulation of the design objective, a piece-wise linear representation of the load deflection curves was used; a number of idealized load-deflection curves that represent existing energy absorbing structural components were developed and are shown in Fig. 3.18.

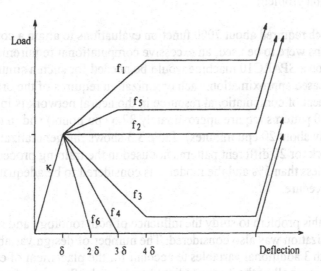

Fig. 3.18. Typical load-deflection curves

A linear combination of these load-deflection curves, with the weighting coefficients designated as design variables, was then used to develop the optimal solution. In addition to the
weighting coefficients, the linear stiffness κ and the linear displacement δ were also designated as design variables. In terms of the individual force-displacements curves f_i, $i=1,2,..6$,
the composite load-deflection curve was written as follows,

$$f = \sum_i c_i f_i \qquad (3.26)$$

where c_i are the weighting coefficients and $f_i=f_i(\kappa,\delta)$. The design problem focussed on the
location of EA structures and their load-deflection behavior in the subfloor region to minimize the accelerations experienced by the pilot and co-pilot. In the context of the KRASH
analysis program, these accelerations were related to a dynamic response index (DRI)
which is indicative of the likelihood of severe spinal injury to the occupants. In the optimization problem, therefore, the DRI was minimized subject to side constraints on the eight
design variables. This can be mathematically stated as follows:

$$\text{Minimize } \overline{F}$$

$$\text{Subject to } \overline{X}^L \leq \overline{X} \leq \overline{X}^U \qquad (3.27)$$

where $\overline{F}$ is a scalar or vector of the DRI values corresponding to different crash conditions,

and is a function of the design variable vector $\overline{X}$; $\overline{X}^L$ and $\overline{X}^U$ are the lower and upper
bounds on the design variables, respectively. A BP network with 8 input nodes (correspond-
ing to the 8 design variables), 5 nodes in the hidden layer, and a single output node (where

the output DRI response was available) was used to approximate the calculations required in the optimization problem.

The genetic search required about 7000 function evaluations to attain a converged design. If no approximations were to be used, an excessive computational requirement of the order of 1750 cpu hours on a SPARC10 machine would be needed for such a simulation. Using the neural network based approximation, each optimization requires of the order of 10 cpu minutes; the investment of computational resource in the neural network is in the generation of training data (150 patterns require approximately 37.5 cpu hours) and in network training (for this case only about 20 cpu minutes). Table 3.3 shows the generalization capability of the neural network for 20 different patterns not used in the training process; all generalization errors were less than 5% and the model was considered to be adequate for use in the optimization procedure.

An extension of this problem to study the influence of both topology and rezing variables in the design optimization was also considered. The number of design variables in this problem were 11, with 3 additional variables to account for the placement of energy absorbing elements under any or all of the 3 longitudinal beams of the floor structure. These topology variables could assume values of 0 or 1, denoting the absence/presence of energy absorbing material under the beam. The BP network in this case consisted of 11 input nodes, 2 hidden layers with 10 and 8 neurons, respectively, and 1 output node. The additional hidden layer in this node was required to make the training more amenable - it appears that the function of the first hidden layer was to partition the mapping into regions based on topology, while the second layer served the usual function of developing the input-output relationship. A total of 300 patterns were used in network training. Using these trained networks, the optimized design was used to compute the DRI value from the program KRASH; this number was only 0.7% different from the network predicted value.

The use of the BP network was also explored in the Stage 2 design problem, where, given the desired load deflection behavior, the geometric characteristics of physical energy absorbing components was established. In previous design studies, a number of energy absorbing components with different material systems, layouts, and geometries have been explored. The analysis of energy absorbing (EA) structural systems requires consideration of the effects of large geometry change, strain-rate, strain-hardening, and various interactions between different collapse modes. Among EA concepts, cylindrical tubes have been extensively studied due to their dual role as energy absorbers and use in normal structural application. Under a compressive load, the wall of a cylindrical tube either buckles into an axisymmetric bellows geometry or a diamond fold pattern. The corresponding load deflection curves exhibit an oscillatory pattern of post-buckling load. While a number of theoretical and experimental studies have been conducted with cylindrical tubes, it is quite clear that simple tubes or simple flat plate structures do not meet requirements for

Table 3.3. Validation of trained neural network

Design #	Neural Network Based Prediction (DRI)	Prediction from KRASH model (DRI)	Error (%)
1	29.98	29.78	0.7
2	33.06	34.60	4.5
3	32.87	32.08	2.5
4	26.76	25.91	3.3
5	34.45	35.71	3.5
6	37.23	37.65	1.1
7	26.67	25.88	3.1
8	28.16	27.22	3.5
9	32.87	31.41	4.6
10	25.91	25.74	0.7
11	33.97	34.24	0.8
12	27.92	27.42	1.8
13	36.71	36.92	0.6
14	25.12	26.19	4.1
15	30.92	29.26	5.7
16	38.01	37.29	1.9
17	27.18	26.36	3.1
18	27.67	27.19	1.8
19	37.87	36.72	3.1
20	24.78	25.14	1.4

an ideal energy absorber; two requirements for such application are a long stroke length and the ability to absorb energy at a uniform rate. Improvements in EA concepts over the conventional straight tube structures are designed to eliminate undesirable characteristics, and the introduction of concertina corrugation is one such approach (Fig. 3.19). In this concept, the utility of the axial collapse of the cylinder is promoted by the formation of plastic hinge mechanisms. Along similar lines, several EA concepts (Fig. 3.20) using corrugated web beams for the aircraft subfloor were proposed by Cronkhite [33]. As a first step in the selection process, ABAQUS finite element models of each concept were used to obtain the load deflection characteristics. The load-deflection characteristics for these structures determined from the ABAQUS static analysis are shown in Fig. 3.21. The corrugated web concepts demonstrated near rectangular load-deflection curves and a high stroke-to-length ratio. It was considered to be the most promising, and the geometry for this component was established to get a desired load-deflection relationship. The significant geometry variables for the problem are the corrugation amplitude 'a', the thickness of the web material 't', and the number of full sine corrugations 'n' in the web.

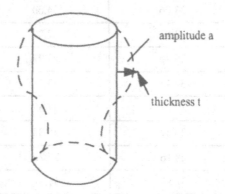

Fig. 3.19. Concertina pattern in a cylindrical shell

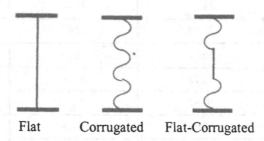

Flat   Corrugated   Flat-Corrugated

Fig. 3.20. Profile of web designs

Parameteric studies were used to establish bounds on the geometry variables as follows.

$$0.32 \leq a \leq 0.62$$
$$0.19 \leq t \leq 0.34 \tag{3.28}$$
$$2 \leq n \leq 4$$

These variables were used to solve an inverse design problem, the objective of which was to choose the dimensions of the corrugated web to produce a load deflection curve that was obtained as an optimal solution to the Stage 1 problem (Fig. 3.22). If $F=[F_1, F_2,...F_n]^T$ represents a vector of loads, where F_i is the load corresponding to the i-th displacement of a load-deflection curve, then the mathematical statement of the optimization problem can be written as follows:

$$\text{Minimize } Z = \sum_n |E_i| = \sum_n |(F_i)_t - (F_i)_a|$$

$$\text{Subject to} \qquad \Delta W \leq \Delta W_{allowable} \tag{3.29}$$

$$X_i^L \leq X_i \leq X_i^U$$

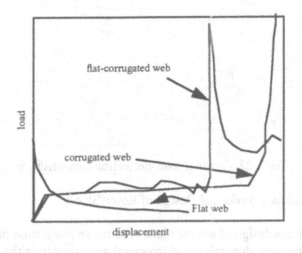

Fig. 3.21. Load-deflection curves obtained from ABAQUS static analysis

Here, the subscripts 't' and 'a' refer to the target and actual values of the loads. The nonlinear load-deflection curve was modeled by a sequence of piecewise linear segments. In order to compute the objective function, which in this problem is the difference between the desired load-deflection curve and that corresponding to a newly produced design, the end points of each linear segment were used as points where this comparison was performed. Since the initial analysis indicated that all the load-deflection curves were flat and rectangular-shaped, a total of three points along the nonlinear region were considered sufficient to represent the load-deflection curve of the corrugated web. The values of loads at these three points corresponded to the values of output nodes in the neural network training. The three points were selected at crush depths of 1 in., 3 in,. and 5 in. Since each FE analysis with ABAQUS required about 10 cpu minutes on a SPARC 10, it was not possible to link the ABAQUS based analysis to the GA search algorithm directly. Instead, a total of 100 exact ABAQUS analyses were executed to generate the training data for the neural network. When tested on data that was not part of the original training set, errors of less than 2% were obtained in predicting the loads for a given displacement. As in the Stage 1 study, considerable savings in computational resource resulted from these approximations. It is worthwhile to point out that the neural network model also works extremely well for discrete/integer variations in the design variables; in this problem, the web thickness and number of sinusoidal half-waves are variables of this type.

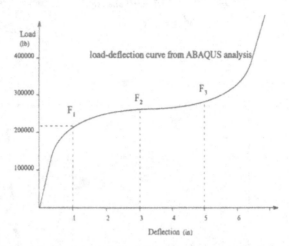

Fig. 3.22. Optimal load-deflection curve from Stage I

3.3.2.3 Multidisciplinary Structural Design of Rotor Blades

The multidisciplinary design of a rotor blade requires an integration of the disciplines of acoustics, aerodynamics, dynamics, and structural analysis within the optimization framework. This simplified example illustrates some of the inherent complexities in such design problems, and the use of neural networks in this context.

The use of composites in rotorcraft blade design provide opportunities for enhanced aerodynamic, structural, and dynamic performance. With composites, it is practical to fabricate non-rectangular blades with variations in twist distribution and airfoil sections along the blade span, thereby contributing to increased flexibility in aerodynamic design. Satisfactory aerodynamic design requires that the required horsepower for all flight conditions not exceed the available horsepower, that the rotor disk must retain lift performance to avoid blade stall, and that the vehicle remain in trim. Important factors in structural design include material strength considerations for both static and dynamic load conditions. A combination of flapwise, inplane, torsion, and centrifugal forces typically comprise the static loading. Another important consideration that encompasses both structural and aerodynamic design, is the autorotation capability. The autorotation requirement pertains to maintaining the mass moment of inertia of the rotor in the rotational plane at an acceptable level. This is a function of the vehicle gross weight, rotor aerodynamic performance, and the rotor system mass moment of inertia. Finally, dynamic design considerations of the rotor blade pertain to the vibratory response of the blade under the applied loads; this design limits the dynamic excitation of the fuselage by reducing the forces and moments transmitted to the fuselage.

A finite element in time and space formulation was used to model the dynamics of the blade [34]. This formulation is based on a multibody representation of flexible structures undergoing large displacements and finite rotations, and requires that the equations of motion be explicitly integrated in time. An unsteady aerodynamic model is used to obtain the induced flow and to calculate the aerodynamic forces and moments in hover and forward flight. In addition to the geometric nonlinearities that are inherent in this problem, the loading on the blade varies as it moves around the azimuth - on the advancing side the flow velocity over the blade is additive to its tangential speed; on the retreating side, these velocities substract. Consequently, in order to maintain force and moment equilibrium, the pitch of the blade is continuously changed as it rotates around the azimuth, and the hub loads are a function of the blade rotational frequency. There is a transient period during which equilibrium of the vehicle is established (trim), and this is generally obtained in 8-10 revolutions of the blade. This process of obtaining trim condition prior to load calculation introduces additional computational costs. The nature of the solution process used here is such that designs that correspond to aeroelastically unstable configurations will simply yield a divergent response in time. As described in a later section, this requires that the neural network based approximation procedure be treated somewhat differently.

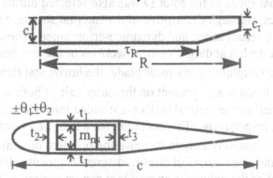

Fig. 3.23a and 3.23b. Planform and profile of rotor blade

Table 3.4 Design variables and bounding value for the rotor design problem

design variable	X	min	max
tuning mass	m^i	0.0	4.0
horizontal flange thickness	t_1^i	0.25	0.35
left vetical flange thickness	t_2^i	0.20	0.30
right vertical flange thickness	t_3^i	0.20	0.30
blade twist [deg]	θ_t	4.0	8.0
twist shape parameter	δ	0.3	1.0
taper inception point	τ_R	0.50	0.80
chord ratio	λ_c	1.0	2.0
rotational speed [rad/sec]	Ω	24.0	30.0
ply angle of inner flange [deg]	θ_1	30	90
ply angle of outer flange [deg]	θ_2	30	90

The objective of the design problem was to design the blade geometry and internal structure to minimize a weighted sum of hub shear force and bending moments for a hingeless rotor blade in forward flight; aerodynamic, performance and structural design requirements are considered as constraints, and dynamic requirements constitute a multicriterion objective function. The premise behind the approach is that a minimization of the hub loads and moments translates into lower vibrations that are transmitted to the fuselage structure.

The planform geometry of the blade is shown in Fig. 3.23a. The blade is tapered in both the chord and the depth, and all loads are assumed to be carried by the structural box shown in Fig. 3.23b. The design variables for this problem are also shown on the figures and summarized in Table 3.4. The blade planform geometry is defined by the chord ratio λ_c and the point of inception of taper along the span denoted by τ_R. The blade twist varies from θ_t at the root to zero at the tip; this variation maybe linear or nonlinear, and is controlled by the twist shape parameter δ. Both θ_t and δ were design variables in the problem. Two nonstructural masses were positioned along the span of the blade, and the magnitude of these masses (m_1 and m_2) as well as their locations along the span (d_{m1} and d_{m2}) were considered as design variables. The rotational speed of the rotor Ω was also selected during the design process. The remaining variables, although descriptive of the internal structure of the blade, have a strong influence on both aerodynamic and dynamic performance. These were the horizontal flange thickness t_1, and the left and right vertical sections of the box beam denoted by t_2 and t_3, respectively. For the Graphite/Epoxy rotor blade, the horizontal flanges are symmetric +/-45 deg laminates; this layup is also present on the outer half of both vertical sections of the box beam. The inner half of the vertical walls are divided into two segments, with a layup of +/-θ_1 and +/-θ_2 deg, respectively. This accounts for a total of 14 design variables for the problem. Note that it is relatively easy to increase the problem dimensionality by simply varying the thickness and orientations of plies in discrete segments along the span. The mathematical statement of the optimization problem can be written as in (3.30).

In this problem statement, F_z, M_y and M_z are the maximum peak-to-peak values of the scaled shear force, flap bending moment, and lead-lag bending moment, respectively; HP_h and HP_f denote the horsepower required during hover and forward flight, respectively, and HP_a is the available horsepower; C_T and σ are the rotor thrust coefficient and solidity, respectively, and the subscripts L and U denote the lower and upper bounds of this ratio. These bounds are necessary to limit the lift performance of the rotor disk to avoid blade stall. The parameter FM is a Figure-of-merit for the helicopter performance out of the ground effect, and subscipt L is used to denote its lower bound; the blade must have sufficient autorotation (AI) capacity to safely descend in an out-of-power configuration, and this is represented in constraint g_7; W is the sum of the structural and nonstructural weights and W_U denotes its upper bound; $\overline{R}$ is used to denote the structural failure criterion, and a Tsai-Wu measure was used for the composite structure in this problem; σ_{all} is the allowable stress for the material and σ_{buck} are the static stresses due to buckling. The constraint set represents response quantities that are traditionally associated with structures, dynamics, and aerodynamic performance of the helicopter. For one given set of design variables, the analysis time required for the evaluation of the objective and constraint functions is substantial (about 18 CPU minutes on a SPARC station) and clearly not amenable to integration with in an iterative optimization environment

$$Minimize \to F = c_1 F_z + c_2 M_y + c_3 M_z$$

Subject · to · constraints

$$g_1 \equiv \frac{HP_h}{HP_a} - 1 \le 0 \qquad g_2 \equiv \frac{HP_f}{HP_a} - 1 \le 0 \qquad g_3 \equiv 1 - \frac{\frac{C_T}{\sigma}}{\left(\frac{C_T}{\sigma}\right)_L} \le 0$$

$$g_4 \equiv \frac{\frac{C_T}{\sigma}}{\left(\frac{C_T}{\sigma}\right)_U} - 1 \le 0 \qquad g_5 \equiv 1 - \frac{FM}{(FM)_L} \le 0 \qquad g_6 \equiv \frac{FM}{(FM)_U} - 1 \le 0$$

$$g_7 \equiv 1 - \frac{AI}{(AI)_L} \le 0 \qquad g_8 \equiv \frac{W}{(W)_U} - 1 \le 0 \qquad g_9 \equiv \bar{R} - 1 \le 0 \qquad g_{10} \equiv \frac{\sigma_{buck}}{\sigma_{all}} - 1 \le 0$$

(3.30)

Table 3.5. Lower/upper bounds on design variables

	design variable	symbol	min	max	accuracy
d_1	mass 1 location	m_1	1	5	1
d_2	mass 2 location	m_2	6	10	1
d_3	tuning mass 1 (kg)	d_{m1}	0.0	5.0	0.1
d_4	tuning mass 2 (kg)	d_{m2}	0.0	5.0	0.1
d_5	horizontal flange thickness ratio to box beam	t_1	0.25	0.35	0.01
d_6	left vertical flange thickness ratio to box beam	t_2	0.20	0.30	0.01
d_7	right vertical flange thickness ratio to box beam	t_3	0.20	0.30	0.01
d_8	blade twist (deg)	θ_t	4.0	8.0	0.1
d_9	twist shape parameter	δ	0.3	1.0	0.1
d_{10}	taper inception point	τ_R	0.50	0.80	0.01
d_{11}	chord ratio	λ_c	1.0	2.0	0.1
d_{12}	rotational speed (rad/sec)	Ω	24	34	1
d_{13}	layup angle of inner vertical flange (deg)	θ_1	30	90	1
d_{14}	layup angle of outer vertical flange (deg)	θ_2	30	90	1

Both the BP and CP neural networks were used to generate the mapping between the design variables and the response quantities of interest. Some of these mappings are quite nonlinear, and require careful consideration of the choice of network architecture and of the number of training patterns. The range of design variable variations used in this training are shown in Table 3.5. A number of training patterns was generated, in the range of design variable variation, and this included both stable and unstable designs. A BP network with 14 input layer neurons, 10 hidden layer neurons, and 5 output layer neurons (a 14-10-5 network), corresponding to 5 output quantities, viz., vertical hub shear, flap moment, lag moment, rotor thrust, and the failure criterion index $\overline{R}$, was established and trained with 550 training patterns. This training presented problems in that it was difficult to reduce the training error to below 3%. The training data was then sorted to separate the designs that yielded stable and unstable responses; it was considered expedient at this stage to establish 5 networks, each mapping the design variables into one output only (14-10-1 networks for all but the lag moment, where a 14-10-8-1 network was used). The training of each of these networks could be done in parallel. The training using these separated patterns procceded well, converging to an error of less than 1%, with the exception of the lag moment (error was 2.6%). The observed pattern of time variation for the lag moment was quite nonlinear, and provides an explanation for the discrepancy in training. Table 3.6 shows the results of testing these networks for generalization performance using 10 sets of design variables that were not part of the original training set. Similar generalization performance was obtained from networks trained with the unstable data. The difficulty in training the networks for the combination of stable and unstable data can be attributed to either a) a completely different input-output relationship in one or more components or b) excessive data for the number of weights and bias constants in the network that could be varied to fit the data.

Table 3.6. Testing of the BP trained network

shear force (N)			flap moment (N m)			lag moment (Nm)			thrust in cruise (N)			failure criterion		
NN	actual	%	NN	actual	%	NN	actual	%	NN	actual	%	NN	actual	%
13695	13546	1.10	11403	11711	2.08	5278	5266	0.24	778	784	0.76	0.261	0.261	0.01
8666	8649	0.19	6295	6891	0.50	4059	3994	1.65	517	506	2.17	0.243	0.241	0.83
6864	6790	1.09	6700	6745	0.67	4036	3880	4.72	369	374	1.33	0.247	0.249	0.84
9160	9373	2.27	8409	8406	0.04	3917	3772	3.84	498	502	0.79	0.313	0.308	1.62
14952	15028	0.51	12905	12984	0.61	4082	4086	0.10	857	873	1.83	0.250	0.250	0.02
17002	17001	0.03	12768	12736	0.25	5974	5899	1.27	926	921	0.54	0.319	0.316	0.94
7284	7339	0.75	6934	6957	0.33	11606	11559	0.40	435	427	1.87	0.238	0.244	2.45
6754	6591	2.47	7542	7569	0.35	8217	7980	2.97	348	352	1.13	0.306	0.307	0.65
13568	13352	1.62	9436	9211	2.45	7365	7104	3.67	722	712	1.40	0.325	0.328	0.91
8652	8671	0.22	8347	8521	2.04	5221	4921	6.08	450	453	0.66	0.235	0.239	1.67

Similar experiments were also performed using the full CP network, that generates an identity mapping of the type [X,Y] -> [X',Y']. This network architecture requires a much larger number of input patterns, and the quality of generalization depends upon the number of Kohonen layer neurons that are permitted (indirectly a measure of maximum cluster radius). If the cluster radius were set to zero, the number of Kohonen neurons will be equal to the number of training patterns, and the network would simply "memorize" all training data. The results of numerical experiments designed to test this network are shown in Table 3.7.

.Table 3.7. Testing of the trained CP network

shear force (N)			flap moment (N m)			lag moment (N m)			thrust in cruise (N)			failure criterion		
NN	actual	%	NN	actual	%	NN	actual	%	NN	actual	%	NN	actual	%
8942	8687	2.93	7316	7504	2.53	3663	3705	1.13	472	483	2.28	0.25	0.258	0.39
14963	14272	4.84	10056	10264	2.02	8273	7892	4.82	766	763	3.93	0.312	0.312	0.01
12384	12579	1.55	10568	10458	1.05	5894	5758	2.36	763	754	1.19	0.235	0.237	0.84
8926	8919	0.08	7045	7058	0.18	5685	5713	0.49	509	502	1.39	0.264	0.266	0.75
10253	10156	0.96	13093	13185	0.69	11796	11183	5.48	933	907	2.86	0.329	0.333	1.20
8113	8046	0.84	8936	8841	1.01	12529	11932	5.00	376	371	1.35	0.347	0.349	0.57
9735	9653	0.85	9618	9685	0.69	22940	23695	3.18	467	454	2.86	0.394	0.390	1.02
14885	14100	5.56	10856	10547	2.93	36633	31888	8.10	650	642	1.25	0.510	0.507	0.59
13883	14016	0.96	10567	10648	0.76	29205	28742	1.61	690	694	0.58	0.412	0.405	1.73
21050	20975	0.36	16889	16205	4.22	29445	30113	2.21	1134	1093	3.76	0.334	0.333	0.31

A BP network was also trained to differentiate between stable and unstable patterns. In the present work, stable patterns were denoted by a 0 and unstable patterns by a 1. The purpose of this network was to act as a filter during optimization to discard the inclusion of unstable patterns in the genetic population. A testing of this BP network showed that it was able to distinguish between stable and unstable patterns correctly 95.4% of the time.

3.3.2.4 Robust Design of Damage Sensing Structures

A "smart" structure, instrumented with sensors and actuators, and responding in an intelligent manner to a dynamically changing environment, is an intriguing concept. The key ingredients in the realization of such a system include an adequate instrumentation of the structure, ability to rapidly analyze measured data and correlate to the existing state of the system, and to limit adverse structural behavior by providing real-time reaction in response to the evaluated state of the system. The strain field in the structure can be an indicator of its overall health, and analytical relations between size and location of damage in composite beams for commonly encountered modes of damage such as delamination, fiber breakage, or matrix cracking are presented in [35] Since the damage can be in more than one place, and furthermore, there can be multiple modes of damage present at the same time, the identification space in such a problem is often nonunique. Artificial neural network based classifiers present themselves as a logical tool for relating specific strain response to damage type and location. Once trained, these networks can rapidly generalize new strain measurements into an estimated state of the structure, and are therefore ideal for online damage detection systems.

An adequate instrumentation of the structure, however, continues to be a pivotal problem. The least number of sensors is clearly desirable from a standpoint of complexity of hardware. However, a sufficient number must be placed to resolve problems of nonunique identification and to have a robust system that is relatively insensitive to partial failures in the sensor array. The problem of optimally locating the least number of sensors that would identify damage over some admissible range of degradation and location is a discrete optimization problem, and its solution requires substantial investment of computational resource. The optimization procedure would require analyses to determine the strain state for many different sizes and locations of damage, and then varying the number and locations of sensors to find the optimal distribution of the sensors in the structure to successfully identify various occurrences of damage. This is clearly a computationally intensive procedure, and in the present work, a trained neural network was used as an approximate analysis tool. Note that in optimizing for sensor locations, each strain reading may correspond to a totally different set of sensor locations. Both BP and CP neural networks were used to construct approximations to the function information required in the optimization process. The approximation was also obtained through a network architecture shown schematically in Fig. 3.24, which is in essence, a combination of the BP and the counterpropagation (CP) networks.

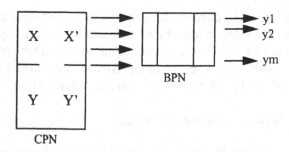

Fig. 3.24. Schematic of a hybrid BP-CP network

The motivation behind examining this architecture was seated in the generally accepted notion that higher quality approximations are generally available from the BP network; however, this network has no capacity for pattern completion. In those instances where generalization of an incomplete input vector is required, the front-end of the proposed hybrid combination (the CP network) can be used to complete the input vector, and which can then presented to the BP network for generalization.

A moderately large number of sensors were first distributed uniformly in the beam structure. Strain readings at these locations under predefined loads, and for a number of random variations in characteristics defining the location and magnitude of damage, comprised the training set for the neural networks described in the previous section. The optimization problem in the present work was formulated to eliminate those sensors from the initial set that were deemed redundant. Such an approach would clearly depend upon the definition of redundancy, and in the present work, those retained sensors were defined as critical for which the error between the network predicted (based on limited strain inputs) and known values of location/extent of delamination corresponding to a set of test strain patterns, was a minimum. This can be mathematically stated as follows:

Given a set of possible sensor locations M by the set S, find $s_i \subset S$ as the positions at which sensors are actually placed so as to minimize

$$F(s) = \sqrt{\frac{1}{P}\sum_{P}\sum_{j=1}^{N}\left(\frac{Y_j^{actual} - Y_j^{predicted}}{Y_j^{actual}}\right)^2} \qquad (3.31)$$

where F(s) is the cumulative error in network prediction for P testing patterns, and N is the number of components used to describe the damage. In the case of delamination in composite beams, N=3 and corresponds to x and z locations of delamination center, and the length of delamination. In this phase of the work, a number of training patterns were generated which assumed the existence of two strain components ε_x and γ_{xz} at each of the M locations of the sensors. The selected neural networks were trained to map these strains into the three parameters describing the delamination size and location.

Note that this optimization problem involves M binary variables, where a 1 or 0 value of the variable indicates whether a sensor is present or absent at a given location, respectively. A solution to this nonlinear integer programming problem is difficult to obtain with traditional mathematical programming algorithms. Consequently, a genetic algorithm based approach was used in this work. The nature of this stochastic search procedure is such that it requires a very significant number of functional evaluations to determine the optimal solution. In the absence of an approximation tool, a new analysis would have to be set up and executed for every trial arrangement of sensors proposed by the optimizer. With trained neural networks available to relate measured strains to the state of the structure, the function evaluations required by genetic search are very inexpensive.

A hygrothermally curvature stable, unsymmetric beam made of Graphite/Epoxy composite laminae of thickness 0.15 mm, was used in this study. The layup sequence was as follows:

$$[30/-60_2/\pm30/60_2/\mp30/-60_2/\pm30/-60_2/-30]$$

The beam was subjected to 4 distinct load conditions as shown in Fig. 3.25 - a distributed transverse load of 1000N/m, a tensile end-load of 5000 N/m, and positive and negative bending moments of 100 N-m and 80 N-m, respectively. Strains corresponding to each of these load conditions were included in constructing the mapping between strains and the delamination characteristics; use of multiple load conditions was important to make the identification space as unique as possible. Assuming that sensors were placed at each of the 15 indicated sites on the beam, 800 sets of training data (strains) were generated in which the x and z location of the delamination center varied between 4-20 cms and 0.15-2.25 mm, respectively. Another set of 15 testing patterns were generated in the range of training data, but were not used for network training. At each sensor location, five strain components were considered to be critical; ε_x, γ_{xz}, due to transverse loading, and ε_x for the tensile load and the two bending moments. This resulted in a total of 75 input strains used to relate to the state of the structure.

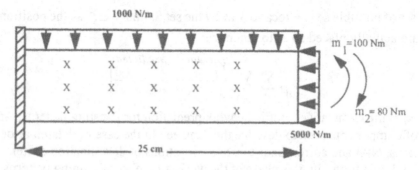

Fig. 3.25. A cantilever beam with 4 distinct loading conditions

A BP network was trained to within 0.01% training accuracy with 200 training patterns. Additionally, the larger training sample with 800 sets of training data were used to train the CP network to comparable levels of resolution. These trained networks were also used to construct the hybrid BP-CP network described above. Each of these networks was tested for its generalization capacity by using the 15 test patterns (strains) not used in training to predict the state of damage in the structure. Strain readings from a number of sensors were eliminated (all strain components from a particular sensor) to simulate the situation of incomplete data, and to assess the generalization capability of each network. Numerical results corresponding to network testing are summarized in Table 3.8, which shows the average error per pattern from each of the three network architectures considered in this work. A total of 15 test patterns were generated randomly in the range of variable variation for which the networks were trained. For the case when all 15 sensor readings are provided at the input, both the BP and CP networks produced minimal errors, with the CP network doing better due to the fact that a larger number of training patterns were used. Strain measurements from the outboard sensors were systematically eliminated from the input vector to assess the quality of approximations available from the network. Quite clearly, the BP network performed the worst, showing no pattern completion capabilities. This poor performance continues to exist even as the number of training patterns are increased.

Table 3.8. Summary of BP and CP network testing

Ns - number of sensors	BP - Average % Error/pattern	CP - Average % Error/pattern
7	126.49	10.04
9	113.67	9.79
12	81.65	4.62
15	2.53	1.36

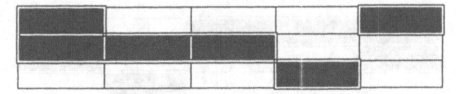

800 test patterns - CPN (100K-N), Ns<7 Ep=29%

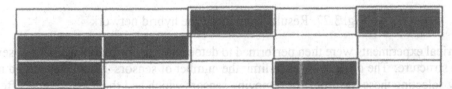

800 test patterns - CPN (400K-N), Ns<7 Ep=11.29%

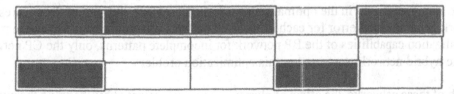

800 test patterns - CPN (603K-N), Ns<7 Ep=11.19%

Fig. 3.26. Results from using CPN network

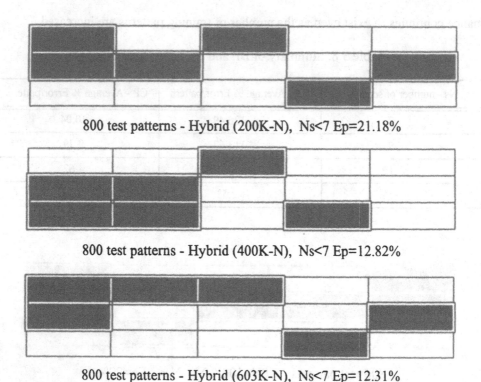

800 test patterns - Hybrid (200K-N), Ns<7 Ep=21.18%

800 test patterns - Hybrid (400K-N), Ns<7 Ep=12.82%

800 test patterns - Hybrid (603K-N), Ns<7 Ep=12.31%

Fig. 3.27. Results from using the hybrid network

Numerical experiments were then performed to determine the optimal placement of sensors on the structure. The objective was to limit the number of sensors below a specified number, by selecting those locations for removing sensors which had the least adverse effect on the generalization capability of the network trained with the original 15 sensors. To implement this example, an additional set of 20 test patterns were developed in the range of network training, and used in the optimal sensor location problem; the sum of the squares of network generalization error for each of these 20 patterns was minimized. Given the poor generalization capabilities of the BP network for incomplete patterns, only the CP network and the hybrid network were used in this optimization problem.

In both of these networks, the effect of cluster size was investigated by changing the number of Kohonen layer neurons in the network. It is important to emphasize the smaller number of Kohonen neurons implies lower data storage requirements, and further, faster look-up capabilities during the network generalization phase. Smaller number of Kohonen neurons also implies a coarser generalization result. Three different cases of cluster size were considered, including 100, 400 and 603 Kohonen neurons. For the number of sensors restricted to less than 7, the results of optimal placements using a CP network are shown in Fig. 3.26. Comparable results of optimal placement using a hybrid network are summarized in Fig.

3.27. In each of these figures, a filled-in area in a particular column or row represents the presence of a sensor as determined by the optimizer. It is obvious from these tables that as the number of Kohonen neurons increases, the generalization capability of both the CP and the hybrid networks improves; in this particular case, increasing the number of Kohonen neurons to above 400 shows only a marginal improvement in the generalization capability. Although results of optimal placements are qualitatively similar for the different cases, there are minor differences which may be attributed to the nature of the stochastic search process.

3.4. NEURAL NETWORKS IN DECOMPOSITION BASED DESIGN

Decomposition based methods have been proposed as a solution to large-scale coupled problems [36,37], wherein the original problem is decomposed into a set of smaller, more tractable subproblems. The coupling between the subproblems may either be hierarchical, where it is possible to identify distinct tree-like patterns of influence (Fig. 3.28), or, no obvious hierarchy may exist and there may be multiple one- or-two-way couplings among the disciplines (Fig. 3.29). The use of decomposition techniques to create a number of smaller subproblems which represent the full complexity of the original problem may allow for parallel processing, and contribute to a better understanding of the problem domain. In general, there are two principal difficulties that have been identified in using decomposition based design techniques. First, in all but the simplest of problems, it is very difficult to establish the topology for decomposition - i.e., assign a subset of design variables and constraints to a particular subsystem. Even when a decomposition topology has been established and smaller subproblems have been created, there is a need to ensure that the coupling between these subproblems is properly taken into the consideration as the design progresses. This section of the chapter describes the use of neural networks to address both of these concerns.

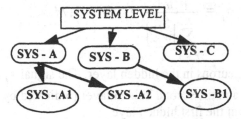

Fig. 3.28. Top-down hierarchical system

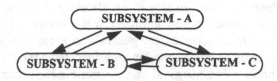

Fig. 3.29. Non-hierarchical system

3.4.1 Analysis of BP Network Weights

In using decomposition principles in multidisciplinary design optimization, an issue that has to be considered at the very outset is the search for a rational approach to decompose the problem. The trained BP network can be used to identify dependencies among design variables and design objectives, and used as a guideline for problem decomposition.

Consider the strengths of the interconnections between the neurons of the various layers of a trained multilayer feedforward network. This matrix of numbers can be viewed as a hologram representing the mapping between the desired input and output quantities, and offers little in terms of a physical interpretation of the problem domain. The flow of information between the input and output layers, however, is a strong function of these interconnection weights, and an analysis of the weights provides meaningful insight into the problem domain.

In order to assess the influence of the i-th input component on the k-th output component, all connecting paths between these neurons must be considered. For each neuron in the hidden layer, the contribution of a signal from the i-th input, $i=1,n$, can be computed as the ratio:

$$s_j = \frac{\left|w_{ji}^{(1)}\right|}{\sum\limits_{r=1}^{n} \left|w_{jr}^{(1)}\right|} \qquad j = 1, M \qquad (3.32)$$

Here M is the number of neurons in the hidden layer. The signal s_j is amplified by the weights between the first and the second hidden layer (indicated by superscript (12)), and summed over all neurons in the first hidden layer,

$$s_k = \sum\limits_{j=1}^{M} \left|w_{jk}^{(12)}\right| \bullet s_j \qquad k = 1, M_2 \qquad (3.33)$$

where M_2 is the number of neurons in the second hidden layer. Finally, the product of these signals with the weights of the output layer is formed and added over all neurons in the second hidden layer as follows:

$$\tilde{t}_{ki} = \sum_{j=1}^{M} \left| w_{jk}^{(2)} \right| \bullet s_j \qquad k = 1, m \qquad (3.34)$$

The quantities in eqn (18) can be normalized to yield elements of the transition matrix $[T]_{nxm}$ as follows.

$$[T] = \frac{\tilde{t}_{ki}}{\sum\limits_{s=1}^{m} \tilde{t}_{ks}} \qquad i = 1, m \qquad k = 1, n \qquad (3.35)$$

For an architecture with a single hidden layer, the above computation simplifies as

$$\tilde{t}_{ki} = \sum_{j=1}^{M} \frac{\left| w_{ji}^{(1)} \right|}{\sum\limits_{r=1}^{n} \left| w_{jr}^{(1)} \right|} \bullet \left| w_{kj}^{(2)} \right| \qquad k = 1, m \qquad (3.36)$$

where the subscripts '1' and '2' denote weights from input to hidden and hidden to output layers, respectively. The elements of the transition matrix show the contribution of the i-th input on the k-th output, and since each row of the matrix sums to unity after normalization, the contribution is available as a fractional quantity. This approach will be referred to as the ABS procedure for weight analysis.

An alternative approach (ALT), and one that preserves the sign of this dependency relationship, is to simply multiply the interconnection weight matrices of each layer in sequence, and normalizing the components by the maximum value in a particular row of the resulting matrix.

$$[\tilde{T}] = \frac{T_{ij}}{max_i |T_{ij}|} \qquad (3.37)$$

where,

$$[T] = \prod_{k=1}^{NL-1} W^k \qquad (3.38)$$

and NL and W^k are the number of layers and the interconnection weight matrices of each layer, respectively.

As an example, consider the 5 bar truss, loaded as shown in Fig. 3.30. If the cross sectional areas of these bars are mapped into the five displacements, the weights can be analyzed (using the ABS procedure) to obtain the data shown in Table 3.9. This data clearly shows the influence of any cross-sectional area on a displacement component. As an example, for the loading considered for this structure, the vertical member is virtually ineffective in influencing the nodal displacements - this is evidenced by the very low numbers in the column corresponding to the cross sectional area A5.

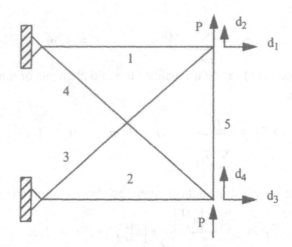

Fig. 3.30. A planar 5-bar truss

Table3.9. Weight analysis for 5-bar truss

Displacements	Areas				
	A1	A2	A3	A4	A5
d1	0.5130	0.1418	0.1330	0.1240	0.0874
d2	0.2670	0.1375	0.3925	0.1832	0.0198
d3	0.1264	0.4655	0.1779	0.2050	0.0252
d4	0.1231	0.2655	0.2588	0.3300	0.0196

The helicopter rotor blade example of the previous discussion provides a more meaningful illustration of the neural network based decomposition procedure. The weights of the trained BP neural networks representing the input-output relationships for this problem were analyzed to establish the topology for decomposition. Both of the strategies for weight analysis were exercised for networks trained with stable and unstable patterns. The resulting transition matrices are shown in Tables 3.10 and 3.11; a qualitative description of the strength of couplings between input and output quantities taken from [38] is shown as Table 3.12. Note that both the ABS and the ALT strategies show in quantitative terms the dominance of specific variables on the 5 outputs considered, and there is qualitative agreement with the assertions made in Table 3.12.

Table 3.10. Transition matrix using the ABS approach

	shear force		flap moment		lag moment		thrust (forward flight)		failure criterion	
	stable	unstable	stable	unstable	stable	unstable	stable	unstable	stable	unstable
d_1	0.0396	0.0551	0.0251	0.0315	0.0227	0.0265	0.0159	0.0284	0.0773	0.0642
d_2	0.0641	0.0697	0.0325	0.0789	0.0853	0.0947	0.0604	0.0478	0.0669	0.0638
d_3	0.0393	0.0644	0.0482	0.0862	0.0304	0.0526	0.0206	0.0679	0.0864	0.1074
d_4	0.0528	0.0184	0.0396	0.0513	0.0673	0.0714	0.0381	0.0675	0.1103	0.0790
d_5	0.0452	0.0374	0.0499	0.0439	0.0436	0.0478	0.0637	0.0503	0.0280	0.0710
d_6	0.0208	0.0393	0.0389	0.0832	0.0445	0.0511	0.0238	0.0396	0.0453	0.0590
d_7	0.0263	0.0922	0.0176	0.0236	0.0476	0.0758	0.0144	0.0635	0.0151	0.0488
d_8	0.2175	0.1342	0.2221	0.1308	0.1137	0.0629	0.2628	0.1402	0.0310	0.0438
d_9	0.1060	0.0842	0.1886	0.1336	0.0947	0.0693	0.1071	0.0913	0.0172	0.0490
d_{10}	0.1062	0.0672	0.0929	0.0565	0.0942	0.0856	0.0827	0.0560	0.1523	0.1130
d_{11}	0.1374	0.1101	0.0728	0.1071	0.0957	0.0850	0.1204	0.1138	0.1776	0.0909
d_{12}	0.1184	0.1501	0.0930	0.1149	0.0638	0.1051	0.0969	0.1584	0.1406	0.1352
d_{13}	0.0067	0.0451	0.0599	0.0346	0.0982	0.0960	0.0511	0.0481	0.0159	0.0436
d_{14}	0.0190	0.0319	0.0182	0.0233	0.0977	0.0756	0.0415	0.0264	0.0356	0.0305

The ALT approach offers additional insights into the direction of change which is unavailable in the ABS approach. These matrices were also obtained for data that was trained with only the unstable samples. The ABS strategy yielded a transition matrix that was not qualitatively different from the one obtained for stable patterns. This may point to the conclusion that there is no significant difference in the input-output relationship between stable and unstable patterns, and that the difficulty in training encountered with a combined set of training patterns is attributable to the network size. If one looks at the transition matrix using the ALT approach, however, there are some differences in signs between dependencies of stable and unstable patterns, and these may very well contribute to training difficulties.

3.4.2. Coordination in Decomposition Based Design

The topology for decomposition in design follows either a hierarchical or non-hierarchical pattern. When the pattern of hierarchy can be established as shown in Fig. 3.28, the solution of the subproblems naturally flows from one level to another, either in a top-down or a bottom-up manner. In problems with multiple one- or two-way couplings between the decomposed subproblems, the system decomposition is said to be non-hierarchical. In non-hierarchical cases, each

	shear force		flap moment		lag moment		thrust (forward flight)		failure criterion	
	stable	unstable	stable	unstable	stable	unstable	stable	unstable	stable	unstable
d_1	0.02074	0.03663	0.00667	0.00880	0.17626	0.06606	0.00957	0.05035	0.26567	0.08070
d_2	0.01966	0.02833	0.01917	0.16332	0.74172	0.42942	0.02640	0.00298	0.24132	0.10599
d_3	0.03711	0.13995	0.01747	0.08013	0.27764	0.25895	0.00027	0.03257	0.12938	0.04444
d_4	0.17700	0.07239	0.09541	0.00472	1.00000	0.56254	0.06866	0.00501	0.63950	0.37660
d_5	0.10597	0.13043	0.16833	0.21025	0.03724	0.00710	0.05728	0.16790	0.08011	0.12244
d_6	0.06228	0.01499	0.00187	0.08546	0.37850	0.07705	0.02465	0.04619	0.15416	0.00583
d_7	0.03873	0.03275	0.00175	0.01133	0.02929	0.04537	0.02913	0.00721	0.05039	0.01807
d_8	1.00000	1.00000	1.00000	1.00000	0.43362	0.11590	1.00000	1.00000	0.11463	0.01481
d_9	0.59041	0.58762	0.67066	0.77645	0.36576	0.54168	0.50425	0.60111	0.05210	0.00199
d_{10}	0.22830	0.07621	0.18173	0.12070	0.11079	0.06268	0.15303	0.12496	0.48158	0.07720
d_{11}	0.34442	0.64685	0.16574	0.42171	0.68886	0.40604	0.19693	0.38843	0.93228	0.64906
d_{12}	0.52357	0.99539	0.34654	0.75995	0.74762	1.00000	0.38188	0.78055	1.00000	1.00000
d_{13}	0.00312	0.06252	0.07220	0.05765	0.19221	0.63768	0.02557	0.03348	0.01729	0.07959
d_{14}	0.00949	0.01208	0.03209	0.01364	0.00116	0.42581	0.03692	0.01530	0.07082	0.00213

Table 3.11. Transition matrix using the ALT approach

variable	Aerodynamics	Dynamics	Structures
planform	S	S	S
twist	S	S	W
tip speed	S	S	S
stiffness	S	S	S
mass distribution	W	S	S

Table 3.12. Qualitative assertion of the strength of coupling

decomposed subproblem has an exclusive subset of design variables and a computed output response. This output response provides the coupling between the subproblems as an output of one subproblem may act as an input to another. A schematic illustration of such a nonhierarchical system is shown in Fig. 3.29. Neural networks have found application in both classes of decomposition driven design, and this is discussed in subsequent sections of this chapter.

3.4.2.1 Hierarchical Systems

In structural optimization, an example of hierarchical decomposition would be one resulting from structural substructuring concepts wherein the entire structure is first analyzed for determining substructure loads. The substructures are then treated as subproblems of the optimization problem, each involving fewer design variables than the global optimization problem. The subproblems are optimized keeping the substructure loads constant. In typical redundant structures, these loads would redistribute with changes in substructure stiffness. This forces a reanalysis of the assembled structure and a repetition of this process to convergence. Since the coupling of substructure stiffness characteristics is not adequately represented in this approach, it is susceptible to generating suboptimal designs.

Sobieski [39] proposed an approach described most appropriately as a linear decomposition strategy. Here, the coupling between subproblems was represented at the coordination problem level; this was achieved by using a linear extrapolation of the subproblem optimal design with changes in the coordination problem design variables. The approach, although effective for the class of problems considered, was not without its drawbacks. Perhaps the most significant problem that could exist with the approach in more realistic problems is the accuracy with which the subproblem optimal design is represented at the coordination problem level. This representation requires the sensitivity of the optimal design in each subproblem to prescribed problem parameters, where the latter are the coordination problem design variables. Methods to obtain this sensitivity are restrictive in that the sensitivity is valid for a limited change in the problem parameters; this translates into tighter move limits in the coordination optimization problem.

Artificial neural network-based approximations can be used to mitigate some of the aforementioned problems. In particular, the multilayered feed-forward network can be used to map the coordination problem design variables into subproblem optimal solutions. This eliminates the need to construct restrictive linear approximations of the subproblem optimal solutions in terms of the coordination problem design variables.

In hierarchical optimization, it is typical to distribute the design variables and design requirements into loosely coupled subgroups. Such a grouping is typically facilitated by classifying the design requirements into either a global or local category. Let the global design variables be denoted by the vector y, and the variables for the i-th subproblem be given as x_i. In terms of these variables, the decomposed optimization problems may be generally stated as follows.

Global or Coordination Problem Statement

$$\underset{i=1}{Minimize} \quad F = F_o(y) + \sum_{i=1}^{NSP} F_i(x_i, y) \tag{3.39}$$

$$Subject \ to \quad g_j(y) \le 0$$

$$g_i(x_i, y) \le 0 \tag{3.40}$$

$$y_k^L \le y_k \le y_k^U$$

Here, the subscript 'o' denotes terms that are strictly dependent on the global variables; similarly, subscript 'i' denotes the influence of the i-th subproblem (i=1,NSP) on the coordination problem. This influence of the i-th subproblem may be regarded as the coupling that exists between the subproblems and the coordination problem. The linking between the various subproblems is only through the global design variables.

Subproblem Statement

$$Minimize \quad F_i(x_i, y) \tag{3.41}$$

$$Subject \ to \quad g_i^S(x_i, y) \le 0 \tag{3.42}$$

$$x_i^L \le x_i \le x_i^U$$

In this subproblem, the design variables are x_i; y may be construed as preassigned problem parameters which are kept fixed during the subproblem optimization. The problem as shown above clearly demonstrates a hierarchy among the design variables - the global variables of the coordination problem are passed through to the subproblems, where they are held constant and treated as preassigned problem parameters. Within the subproblem, the local variables are the only active variables. Similarly, only the global variables are active at the coordination problem level. However, if the sensitivity of the subproblem optimal solution with respect to the global variables is calculated in each subsystem, then the subproblem optimal solution can be approximated at the coordination problem level as follows:

$$x_i = x_i^* + \left(\frac{dx_i^*}{dy}\right)^T \Delta y$$

$$\tag{3.43}$$

$$F_i = F_i^* + \left(\frac{dF_i^*}{dy}\right)^T \Delta y$$

The methods of computing the sensitivity of subproblem optimal solutions are not well developed, and for the above linear approximation to be valid, very restrictive move limits on the global variables have to be imposed at the coordination problem level. Instead, a BP network can be used to develop the nonlinear relationship between the subproblem optimal solution and the global design variables. Such an approach circumvents the need to construct the linear approximation required at the global level. Consider the following design example related to the design of a three element portal frame structure shown in Fig. 3.31 for minimal weight, with constraints on the six nodal displacements and rotations, and on the bending stresses in each element. The cross section of the beams is an I-section and the structural weight and nodal displacements can be defined in terms of the three cross sectional areas and three section moments of inertia.

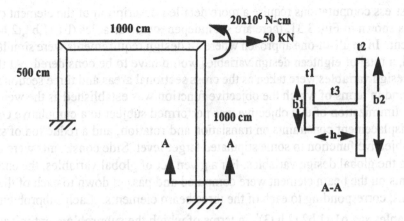

Fig. 3.31. A 3-element portal frame

Table 3.13. Solutions for the portal frame problem

Element ID	Design Variable cm	One-level Displacement and Buckling Constraints	One-level Displacement, stress and Buckling Constraints	Two-level Displacement, stress and Buckling Constraints
1	b1	9.591	8.169	7.883
1	t1	0.494	0.51	0.476
1	h	77.023	75.263	78.412
1	t3	0.342	0.334	0.435
1	b2	8.445	8.109	8.461
1	t2	0.564	0.514	0.492
2	b1	5.119	3.962	3.771
2	t1	0.234	0.246	0.313
2	h	0.01	6.364	5.790
2	t3	0.01	0.331	0.401
2	b2	3.939	3.854	4.320
2	t2	0.273	0.281	0.323
3	b1	7.895	11.461	10.032
3	t1	0.434	0.907	0.868
3	h	71.093	71.368	73.227
3	t3	0.315	0.317	0.361
3	b2	10.723	18.587	15.312
3	t2	0.369	0.552	0.616
Volume (cm^3)		49,972.2	64,178.9	70,472.7

However, stress computations require a more detailed description of the element cross section, and as shown in Fig. 3.31, there are six indepenedent variables (b1,t1,b2,t2,h,t3) for each element. In an all-in-one approach where all design requirements were simultaneously considered, a total of eighteen design variables would have to be considered. At the global level, the design variables were taken as the cross sectional areas and three section moments of inertia, and in terms of which the objective function was established as the weight of the structure. Minimization of this objective was performed subject to a cumulative constraint of all six displacement constraints on translation and rotation, and a reduction of subproblem level objective function to some stipulated target level. Side constraints were also imposed on the global design variables. For a given set of global variables, the end loads and moments on the beam element were computed and passed down to each of the three subproblems, corresponding to each of the three beam elements. Each subproblem has six design variables $x^i=(b1,t1,b2,t2,h,t3)^i$, in terms of which the subproblem optimization problem was formulated. The objective function at the subproblem level was the minimization of the stress constraints at the end location of the beams and a simultaneous minimization of the square of the difference between the values of coordination level variables A_i and I_i passed down from the coordination level, and the computed values of these parameters at the subsystem level. Constraints at this level were imposed against local buckling in the flange and webs of each beam element and side constraints were also placed on the subproblem design variables

The two level approach was then implemented, using the neural network to represent the coupling between coordination and subsystem levels. There were three subsystems in this problem, each defined by six design variables (section dimensions) and two problem parameters (sectional area Ai and moment of inertia Ii). Four networks were trained as follows. The first mapped six areas and moments of inertia into one cumulative measure of displacement constraint. This network was used in lieu of exact analysis at the coordination problem level. Each of the other three networks was used to map six section dimensions into a measure of normal bending stress constraint for the corresponding element. Three other networks were obtained to represent coordination and subsystem level couplings at the coordination level. Each of these networks mapped two coordination level variables into the corresponding subsystem variables and the subsystem level objective function. They were then used to obtain a predict the value of the subproblem at the coordination level. Results from this two level approach are summarized in Table 3.13 and compared against results obtained from a non-decomposition based approach.

3.4.2.2 Non-Hierarchical Systems

As stated earlier, the decomposition approach breaks down the non-hierarchically coupled optimization problem into a number of subspaces, each with a unique set of design variables. An approach for solving such optimization problems is the concurrent subspace optimization (CSSO) strategy. In this approach, each subspace has its set of variables and can directly compute state variables or outputs with these variables. Non-local states must be

approximated within each subspace. The approach uses a cumulative representative constraint to represent all constraints for the subspace. Following design variable allocation,
temporarily decoupled subspace optimizations are performed in each subspace concurrently. The goal of these subspace optimizations (SSO) is to reduce the violation of the
cumulative constraint with the least increase of the system objective function or greatest
decrease if the cumulative constraint is already satisfied. The optimization for the k-th subspace can be stated as follows,

$$Min \quad f(x)$$

$$St \quad C^p \le C^{po}[s^p(1-r_k^p)+(1-s^p)t_k^p] \quad p = 1...nss \tag{3.44}$$

$$x_L^k \le x^k \le x_U^k$$

where, C^P is a measure of all constraints of subspace p; superscript p and k denote the influence of the p-th subspace on the k-th subspace. The $r_k{}^P$ coefficient represents the responsibility assigned to the k-th subspace for reducing the constraint violation of the p-th
subspace. The $t_k{}^P$ coefficient represents the trade-off associated with each subspace that
allows for the violation of a constraint in the p-th subspace, provided a corresponding compensation is south in the k-th subspace. The switch parameter s^P simply enables or disables
the responsibility or trade-off coefficients.

The constrained optimum resulting from these subspace optimizations is a function of the r
and t coefficients, an optimum sensitivity analysis (OSA) is performed in order to determine
the sensitivity of F to these coefficients. This derivatives obtained in the OSA are then used
in the coordination optimization problem (COP) in which the system objective function is
minimized with respect to r and t coefficients. The result of the COP execution is a new set
of r's and t's to be used in the next SSO's. The coordination problem is defined as follows.

$$Min \quad F = F^o + \sum_p \sum_k \frac{df}{dr_k} \Delta r_k^p + \sum_p \sum_k \frac{df}{dt_k} \Delta t_k^p$$

$$St \quad \sum_k r_k^p = 1 \quad p,k = 1...nss$$

$$\sum_k t_k^p = 0 \tag{3.45}$$

$$0 \le r_k^p \le 1$$

$$r_{kl}^p \le r_k^p \le r_{ku}^p \quad and \quad t_{kl}^p \le t_k^p \le t_{ku}^p$$

Following the update of the coefficients, the entire process is repeated until ultimate convergence requirements are met. As in the hierarchical decomposition problem, the fundamental
difficulty is in the reliability of the optimal subspace solution sensitivity to the responsibility
and trade-off coefficients, and which is used to construct approximations in the coordination
problem.

The principal advantages for using artificial neural networks in this application are two-
fold. At the very outset, they can be used to create approximations to typical nonlinear rela-

tions between the design variables and the objective/constraint functions. This is extremely useful in cases where such information can only be obtained as an outcome of expensive function analysis, or in other cases, from an experimental process. The method circumvents the need to calculate gradient information that may not only be expensive, inaccurate, or simply unavailable as may be the case when discrete/integer variables are involved. A second important use of neural networks is in creating approximations to relations that are required to coordinate the solution among the many subsystems. In many cases, these relations may be highly nonlinear and not amenable to first order approximation methods that are typically used; artificial neural networks provide an effective alternative to represent such information more effectively. The work here describes the use of neural networks in developing new coordination strategies [40,41].

To develop such approximations using neural networks, one needs to generate data within the prescribed domain through uniform random distribution, a biased random distribution, or using a design of experiments approach. The following mapping can be generated through the use of either BP or CP networks,

$$\{x_1, x_2, ... x_n\} \Rightarrow \{F, C_A, C_B, ... C_{NSS}\} \tag{3.46}$$

where system variables are mapped to system objective function and cumulative constraints corresponding to all contributing subsystems. In any phase of the subsystem optimization, when design variables are presented to the artificial neural network, it produces a corresponding set of output quantities which are ultimately essential in each subspace optimization. This capability of the neural networks eliminates the need for either full function analysis or use of linear approximations based on coupled system sensitivity analysis; the mapping also eliminates the need to create linear approximations of one subproblem behavior for use in another subproblem. This kind of function approximations are expected to yield good results to the degree of representation (or generalization performance) of the artificial neural networks.

Another approximation that is useful in implementing the CSSO approach is the relation between optimal values (x_A^*, x_B^*, F_A^*, F_B^* and F^*) obtained from local subsystem optimization, and the choice of responsibility and trade-off coefficients for which the optimal values were obtained. Here, x_A^* and x_B^* are the optimal values of design variables in subsystems A and B, respectively. During the optimization process in subsystem A for example, x_B is held fixed at the starting value as is x_A in subsystem B. While F_A^* and F_B^* are the optimal values of the system level objective function obtained in each subsystem, F^* is the computed value of the system level objective function obtained by using the design variable vector x as a concatenation of x_A^* and x_B^*. The coordination procedure is required to guide the solution in such a manner that F^*, and F_A^* and F_B^* all converge to the same point. The principal gain in using neural network approximations is that the knowledge from the computational domain, obtained from a number of exploratory analyses conducted at some randomly (or rationally) distributed points, can be used to guide the solution towards a converged solution. The CP network can be used in two different strategies for coordination purposes.

In the first case the trained CP network is obtained as follows. For a number of different s, r, and t coefficient combinations, several subsystem optimal solutions were obtained and then used as training data to generate a mapping of the type,

$$\{x^*, s_p, r, t\} \Rightarrow \{F^*, F_A^*, F_B^*\} \tag{47}$$

where x^* is obtained by concatenating x_A^* and x_B^*. This CP network is then used in the solution of the coordination problem, which uses the responsibility and trade-off coefficients as design variables to minimize a modified objective function that penalizes any difference between the objective function values obtained in each subproblem optimization as follows:

$$\bar{F}^*(r, t) = F^* + R_{pen}(F_A^* - F_B^*)^2 \tag{48}$$

In this problem, R_{pen} is a penalty parameter, and the optimal values of r and t coefficients obtained as a solution are then used in the next round of subsystem optimization. In eqn (47), the difference between two subsystem objective functions are minimized by adjusting the design variables r and t to enforce the overall system convergence. The use of the CP network in this manner precludes the requirement of computing the sensitivity of system objective function with respect to the r and t coefficients, and using this sensitivity to construct linear approximations of objective function with respect to the r and t coefficients for use in the coordination problem.

An alternative approach has also been examined that altogether eliminates the need for COP solution, and requires a network trained to yield the following relationship.

$$\{F_A^*, F_B^*, x_A^*, x_B^*\} \Rightarrow \{s_p, r, t\} \tag{49}$$

Here, the optimal value of subproblem objective functions and design variables are mapped onto s_p, r and t variables. The subsystem optimization solutions are then initiated by setting all s, r, and t coefficients to zero. These subsystem solutions are then used to predict new values of the s, r, and t coefficients, and for this selection, the optimization in each subsystem is repeated. This procedure is recursively invoked until no further change in the subsystem solutions is realized. In this approach, there is no explicit attempt to drive the solution of the subsystem optimal solutions to a common converged point. A heuristic explanation of why the approach should work in the absence of a coordination scheme is as follows. The training of the network of eqn. (48) requires that to obtain the subsystem optimal solutions, not only must the r, s, and t coefficients be specified, but also a starting vector of design variables. The subsystem optimal solutions, therefore, are a function of the global design variable vector. This relationship is implicitly reflected in the mapping of eqn. (48). Additional safeguards against divergent behavior are built in through specification of move-limits on design variables within each subsystem.

It is important to note that in either approach, the BP or CP network based mapping of eqn. (45) can be used to alleviate the computational requirement involved with obtaining the sub-system optimal solutions used for training the neural networks described by eqns. (46) and (48). The following sections demonstrate the implementation of the proposed ideas in a highly non-linear, internally-coupled design problem.

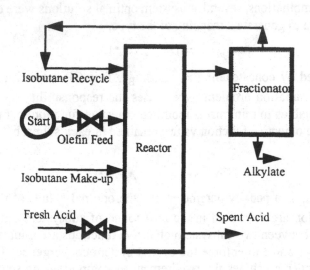

Fig. 3.32. Schematic sketch of the alkylation process

The following coupled optimization problem dealing with an alkylation process (shown in Fig. 3.32) was used to test the aforementioned procedures. The optimization problem exhibits complex state coupling in that an iterative solution procedure is required to solve for both the objective and constraint functions. Fig. 3.32 shows a reactor into which olefin feed and isobutane make-up are introduced. Fresh acid is added to catalyze the reaction and spent acid is withdrawn. The hydrocarbon product from reactor is fed to a fractionator, and isobutane is taken from the top and recycled back to the reactor. Alkylate product is withdrawn from the bottom of the fractionator. The problem includes three design variables and eight states, and the process is constrained by fourteen inequality constraints. Details of the design variables and state variables are beyond the purview of this discussion but may be obtained in [42]. The objective of the profit maximization problem may be stated as follows,

$$y_1(x) = 0.063 y_2 y_5 - 5.04 x_1 - 3.36 y_3 - 0.035 x_2 - 10 x_3 \tag{3.50}$$

where the variable bounds and constraints are presented in Table 3.14

.Table 3.14. Bounds on decision and state variables

Design variable bounds
$0.0 \leq x_1 \leq 2000.$
$0.0 \leq x_2 \leq 16000.$
$0.0 \leq x_3 \leq 120.$

Objective and constraints
y_1 = objective function
$0.0 \leq y_2 \leq 5000.$
$0.0 \leq y_3 \leq 2000.$
$85.0 \leq y_4 \leq 93.$
$90.0 \leq y_5 \leq 95.$
$3.00 \leq y_6 \leq 12.$
$0.01 \leq y_7 \leq 4.0.$
$145. \leq y_8 \leq 162.$

Decomposition into two subsystems A and B can be written as shown below.

$$\{x_A\}^T = \{x_1, x_3\}$$
$$\{y_A\}^T = \{y_1, y_2, y_3, y_4\} \tag{51}$$
$$\{x_B\}^T = \{x_2\}$$
$$\{y_B\}^T = \{y_1, y_5, y_6, y_7, y_8\}$$

This decomposition results in a coupled system in which the state solution requires an iterative approach. In particular, the process optimization objective function and all states (constraints) y_2 through y_8 are highly nonlinear functions of the design variables. The non-
hierarchic structure of the subsystem couplings are shown in Fig. 3.33. The problem was
solved using a non-decomposition based approach, a traditional implementation of the
CSSO approach, and strategies described in previous sections that use neural networks to
enhance the CSSO approach. First, trained BP and CP networks were used for function
approximation in the CSSO procedure. A four layer architecture (3*13*7*3) was selected
for the BP network. Both networks were trained three different times, using n=500,350 and
200 training patterns, respectively. In general, the representation of the mapping between
input-output variables gets progressively degraded as the number of training patterns used
to develop the CP and BP networks are reduced. The training data was generated by varying
the design variables obtained from the all-in-one approach in a uniform random manner in a
range of ±35 % for each design variable (Case 1). As a second step, n=200 data was generated by using design of experiments approach (Case 2). For each of the cases above, maximum generalization errors obtained for testing patterns that were not part of the original
training set, were less than 3%. Cases 3a and 3b in Table 3.15 correspond to using the CP
network approximations of eqns. (46) and (48), respectively.

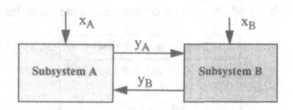

Fig. 3.33. A 2-way coupling between subsystems

As shown in Table 3.15, the results obtained in CSSO approach compare well with those obtained when using neural networks for function approximation. The number of system analysis required in the all-in-one and CSSO approach are shown in Table 3.16. When neural networks are used in the CSSO approach, the number of system analyses are driven by the amount of training data required by the neural networks. This is clearly problem dependent. In using a traditional CSSO which require sensitivity computations, not only are there problems in computing sensitivities that are associated with high dimensionality and matrix inversion, but the number of system analyses increase significantly with the number of iterations, design variables and states variables.

.Table 3.15. CSSO results using neural network approximations

Case	Method	Objective	x_1	x_2	x_3
1	BP(n=500)	-1161.59	1727.21	16000.	97.84
	BP(n=350)	-1160.74	1725.01	16000.	98.20
	BP(n=200)	-1159.86	1726.74	16000.	95.25
	CP(n=500)	-1160.85	1725.24	16000.	97.94
	CP(n=350)	-1157.14	1721.15	16000.	94.38
	CP(n=200)	-1154.73	1725.05	16000.	92.56
2	BP(n=200)	-1157.39	1718.75	16000.	95.26
	CP(n=200)	-1159.84	1722.82	16000.	98.35
3a& 3b	CP	-1160.13	1723.36	16000.	97.85
	CP	-1160.54	1724.43	16000.	97.78

The design problem was also solved using the CP network based coordination strategies described in eqns. (46) and (48), respectively; this network was trained using 200 training patterns generated in a uniform random manner. For this particular problem, the starting solution was feasible, and subsequent moves in design variables were also through the feasible domain. Consequently, throughout the optimization process only the t variables were required to be active. For the design domain of interest, t variables were generated between their allowed range of -1 to +1, as specified in the original CSSO approach. The s and r variables were eliminated from the mappings of eqns. (46) and (48). In both cases, testing error for the CP network was reduced to less than 8% before using in the optimization process. The results for these numerical experiments are shown in Table 3.15, and compare well with results from the CSSO approach. Case a relates to imposing a 10% move limit on design variables while case (b) allows a move limit of 20%. In this highly nonlinear problem, it is quite apparent that the larger move limit leads to a degradation in the quality of the results.

Table 3.16. Results for the one-level and CSSO approaches

One Level Opt.	CSSO (a)	CSSO (b)
Obj = -1162.04	Obj = -1161.96	Obj=-1141.38
x_1 = 1728.38	x_1 =1728.35	x_1=1687.92
x_2 = 16000.0	x_2 = 16000.0	x_2 =16000.0
x_3 = 98.186	x_3 = 97.54	x_3 = 91.86
SA: 135	SA: 82	SA: 82

CLOSING REMARKS

This chapter provides an introduction to four different neural network architectures - the BP network, the modified CP network, the ART network, and the Hopfield network. The basic strategy for training these networks is discussed, including a review of their limitations. The chapter focusses on the applications of these network architectures in problems of structural analysis and design. The ART network is shown to be applicable as a tool for pattern classification, such as may be needed in problems of preliminary design. The Hopfield network can also be used for vector classification; however, more significant applications of this network have been in problems of combinatorial structural optimization. One such problem is discussed in this chapter. The BP and CP network are shown to be effective means of constructing function approximation. Specific applications of such function approximations that have received attention in this work include approximate response evaluation, in the design and functioning of damage sensing monitored structures, and in decomposition based design of structural systems. Both hierarchical and nonhierarchical decomposition topologies are discussed in this context.

REFERENCES

1. Rumelhart, D.E. and NcClelland J.L..: Parallel Distributed Processing, Volume 1, The MIT Press, Cambridge, Massachussets, 1988.

2. Rumelhart, D.E. and NcClelland J.L..: Parallel Distributed Processing, Volume 2, The MIT Press, Cambridge, Massachussets, 1988.

3. Hajela, P..: Stochastic Search in Discrete Structural Optimization - Simulated Annealing, Genetic Algorithms and Neural Networks, Discrete Structural Optimization, Springer, New York, pp. 55-134, (ed. W. Gutkowski), 1997.

4. Neter, J., Wasserman, W., and Kutner, M.H..: Applied Linear Regression Models, 2nd ed., Richard D. Irwin, Inc., 1989.

5. Hajela, P. and Kim, B..: "Classifier Systems for Enhancing Neural Network Based Global Function Approximations", proceedings of the 7th AIAA/NASA/ISSMO/USAF Multidisciplinary Analysis and Optimization Meeting, St. Louis Missouei, 1998.

6. Barron, A.R..: "Neural Network Approximation", proceedings of the Seventh Yale Workshop on Adaptive and Learning Systems, pp. 69-72, Yale University, New Haven, CT, 1992.

7. Hecht-Nielsen, R..: "Counterpropagation Networks", Journal of Applied Optics, Vol. 26, 1987, pp. 4979-84.

8. Szewczyk, Z., and Hajela, P..: "Feature Sensitive Neural Networks in Structural Response Estimation", proceedings of the ANNIE'92, Artificial Neural Networks in Engineering Conference, November 1992.

9. Fu, B. and Hajela, P..: "Minimizing Distortion in Truss Structures: A Hopfield Network Solution", Computing Systems in Engineering, vol. 4, no. 1, 69-74, 1993.

10. Carpenter, G.A. and Grossberg, S..: "A MAssively Parallel Architecture for a Self-Organizing Neural Pattern Recognition MAchine", Computer Vision, Graphics, and Image Processing, Vol. 37, pp. 54, 1987.

11. Grossberg, S..: (ed.), The Adaptive Brain, Vol. I and II, Amsterdam, North-Holland, Elsevier, 1987.

12. Grossberg, S..: (ed.), Neural Networks and Neural Intelligence, Cambridge, MA, MIT Press, 1988.

13. Fu, B. and Hajela, P., and Berke, L..: "ART Networks in Automated Conceptual Design of Structural Systems",   Computing Systems in Engineering, Vol. 4, No. 2-3, pp.121-133, 1993.

14. Sobieszczanski-Sobieski, J..: "Multidisciplinary Design Optimization: An Emerging New Engineering Discipline", World Congress on Optimal Design of Structural Systems, Rio de Janeiro, Brazil, August 2-6, 1993.

15. Tolson, R.H. and Sobieszczanski-Sobieski, J..: "Multidisciplinary Analysis and Synthesis: Needs and Opportunities", AIAA Paper No. 85-0584, 1985.

16. Abdi, F., Ide, H., Levine, M., and Austel, L..: "The Art of Spacecraft Design: A Multidisciplinary Challenge", 2nd NASA/Air Force Symposium on Recent Advances in Multidisciplinary Analysis and Optimization, NASA CP-3031, Sep. 1988.

17. Venter, G. et. al..: "Construction of Response Surfaces for Design Optimization Applications," proceedings of the 6th AIAA/NASA/USAF/ISSMO Conference on Multidisciplinary Analysis and Optimization, pp.548-564, September 4-6, 1996, Bellevue, Washington.

18. Wang, B..: "A New Method for Dual Response Surface Optimization", proceedings of the 6th AIAA/NASA/USAF/ISSMO Conference on Multidisciplinary Analysis and Optimization, pp.1805-1814, September 4-6, 1996, Bellevue, Washington.

19. Giunta, A.A..: "Aircraft Multidisciplinary Design Optimization Using Design of Experiments Theory and Response Surface Modeling," Ph.D. dissertation, Virginia Polytechnic Institute and State University, May, 1997.

20. Hajela, P. and Berke, L..: "Neurobiological Computational Models in Structural Analysis and Design", Computers and Structures, Vol. 41, No. 4, pp. 657-667, 1991.

21. Berke, L., and Hajela, P..: "Application of Artificial Neural Networks in Structural Mechanics", Structural Optimization, Vol 3, No. 1, 1992.

22. Berke, L., and Hajela, P..: "Application of Artificial Neural Networks in Structural Mechanics", NASA TM-102420, 1990.

23. Alam, J., and Berke, L..: "Application of Artificial Neural Networks in Nonlinear Analysis of Trusses", NASA TM, 1993.

24. Ghaboussi, J., Garrett, J.H., Jr., and Wu, X..: "Knowledge-Based Modeling of Material Behavior with Neural Networks", Journal of Engineering Mechanics, 117 (1), 1991, pp. 132-153.

25. Brown, D.A., Murthy, P.L.N., and Berke, L..: "Computational Simulation of Composite Ply Micromechanics Using Artificial Neural Networks", Microcomputers in Civil Engineering, 6, 1991, pp. 87-97.

26. Szewczyk, Z., and Hajela, P..: "Feature Sensitive Neural Networks in Structural Response Estimation", proceedings of the ANNIE'92, Artificial Neural Networks in Engineering Conference, November 1992.

27. Szewczyk, Z., and Hajela, P..: "Neural Network Based Selection of Dynamic System Parameters", Transactions of the CSME , Vol. 17, No. 4A, pp. 567-584, 1993.

28. Szewczyk, Z., and Hajela, P..: "Neural Network Based Damage Detection in Structures", proceedings of the ASCE 8th Computing in Civil Engineering Conference, Dallas, Texas, June 6-8, 1992.

29. Soeiro, F.J..: Structural Damage Assessment Using Identification Techniques, Ph. D. dissertation, University of Florida, 1990.

30. Hajela, P., and Lee, E..: "Topological Optimization of Rotorcraft Subfloor Structures for Crashworthiness Considerations", Computers and Structures, vol. 64, no 1-4, pp. 65-76, 1997.

31. ABAQUS User's Manual, Vol. I, Version 5.2, 1992.

32. Wittlin, G. and Gamon, M.A..: Experimental Program for the Development of Improved Helicopter Structural Crashworthiness Analytical and Design Techniques, USAAMRDL Technical Report 72-72A,72Bi, May 1973.

33. Cronkhite, J.D. and Berry, V.L..: Crashworthy Airframe Design Concepts - Fabrication and Testing, NASA CR-3603, National Aeronautics and Space Administration, Washington, DC, September, 1982.

34. Bauchau, O. A. , and Kang, N.K..: "A Multibody Formulation for Helicopter Structural Dynamic Analysis", Journal of American Helicopter Society, Vol. 38, No. 2, pp. 3-14, April 1993.

35. Teboub, Y..: Integrated Design of Composite Structures for Damage Detection and Mitigation, Ph.D. Thesis, Rensselaer Polytechnic Institute, Troy, New York, 1996.

36. Sobieski-Sobieszczanski, J..: "Optimization by Decomposition: A Step from Hierarchic to Non-Hierarchic Systems. In Recent Advances in Multidisciplinary Analysis and Optimization", NASA CP 3031, 1988.

37. Bloebaum, C. L., Hajela, P. and Sobieski-Sobieszczanski, J..: "Non-Hierarchic System Decomposition in Structural Optimization", Engineering Optimization, Vol. 19, pp. 171-186, 1992.

38. Adelman, H., and Mantay, W.A..: (eds), Integrated Multidisciplinary Optimization of Rotorcraft: A Plan for Development, NASA TM 101617, May 1989.

39. Sobieski-Sobieszczanski, J..: "A Linear Decomposition Method for Large Optimization Problems-Blueprint for Development", NASA TM 83248, February 1982.

40. M. Arslan and P. Hajela.: "Counterpropagation Neural Networks in Decomposition Based Optimal Design", Computers and Structures, vol 65, no. 5, pp. 641-650, December 1997.

41. Arslan, M.A., Hajela, P..: "Use of Artifical Neural Networks to Enhance the Concurrent Subspace Optimization Strategy", proceedings of the International Symposium on Optimization and Innovative Design, JSME Paper No. 153, edited by H. Yamakawa, M. Yoshimura, S. Morishita and M. Arakawa, July 28-July 30, 1997, Tokyo, Japan.

42. Arslan, M.A..: Domain Decomposition in Multidisciplinary Design: Role of Artificial Neural Networks and Intelligent Agents, Ph.D dissertation, Rensselaer Polytechnic Institute, April 1998.

CHAPTER 4

THE NEURAL NETWORK APPROACH IN PLASTICITY AND FRACTURE MECHANICS

P.D. Panagiotopoulos[†]
Aristotle University, Thessaloniki, Greece

Z. Waszczyszyn
Cracow University of Technology, Cracow, Poland

ABSTRACT

The Sections devoted to the applications of neural networks in plasticity and fracture mechanics cover three topics. The first one is associated with the implementation of hybrid programs in which neural procedures are used for the analysis of elastoplastic constitutive equations by means of back-propagation neural networks. The first program corresponds to the bending analysis of elastoplastic beams. The second program deals with the analysis of elastoplastic plane stress problem. The second topic is related to the so-called *Panagiotopoulos approach*. The approach depends on the formulation of the Quadratic Programming Problems and then analyzing them by the Hopfield-Tank network. This approach was used successfully for the analysis of unconstrained and constrained QPPs associated with the classical crack problem and the analysis of elastoplastic structures. The third topic corresponds to the parameter identification problem. This problem is analyzed by means of two neural networks. The supervised learning of a simple backpropagation neural network interacts with the analysis of subsidiary equations by means of the Hopfield-Tank network.

4.1. HYBRID NEURAL-NETWORK/COMPUTATIONAL ANALYSIS OF ELASTO-PLASTIC STRUCTURES

4.1.1. Neural networks and neural procedures

The backpropagation neural network can be used as a quick simulator to map input to output data for complex relations between them. Such a network is trained off-line and as a neural procedure it can then be incorporated into a computer program instead of the corresponding numerical procedure. This leads to hybrid, neural-network/computational strategies and programs [1,2].

Neural procedures associated with BP neural networks can be applied to the analysis of constitutive equations. The BPNNs were earlier used to formulate the stress-strain relations in concrete [3]. The moment-curvature relation was established on the base of experimental data [4] or analytical formulae [5]. The invertion of uniaxial Ramberg-Osgood relation was performed in [6].

The idea of implementation of neural procedures in the finite difference or finite element programs was led in [5,7]. The elastoplastic plane problem and bending of elastoplastic plates were analyzed by hybrid NN/FEM programs in [8,9].

In the following Section two from the above mentioned problems are discussed. The problems concern the applications of neural procedures to the analysis of elastoplastic structures, especially the bending of beams and plane stress problem.

4.1.2. Bending of elastoplastic beams

4.1.2.1. Basic equations

Linear geometric and equilibrium equations are assumed:

$$v' = \varphi(x), \quad \varphi' = -\kappa(x), \quad T' = -p(x), \quad M' = T(x), \tag{4.1}$$

where $(\cdot)' = d(\cdot)/dx$ and other variables are shown in Figs.4.1a,b.

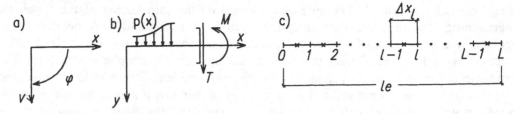

Fig.4.1: a) Generalised displacements, b) Load and cross-sectional forces, c) Finite difference nodes along the beam axis

Eqs (4.1) have to be completed by physical relationship:

$$\kappa = \begin{cases} M/(EI) & \text{for elastic deformation,} \\ f(M) & \text{for elastoplastic deformation.} \end{cases} \tag{4.2}$$

In a general case the elastoplastic relationship $\kappa = f(M)$ cannot be formulated explicitly. This relationship depends not only on material model $\sigma(\varepsilon)$ but also on the shape of beam cross-section.

In what follows a special case is discussed corresponding to the following assumptions: 1) beam is of rectangular cross-section, 2) material is of symmetric properties for tension/compression, with linear strain-hardening and perfect Bauschinger's effect (Fig.4.2a), 3) Bernoulli-Euler hypothesis of plane, perpendicular cross-section obeys (Fig.4.2b).

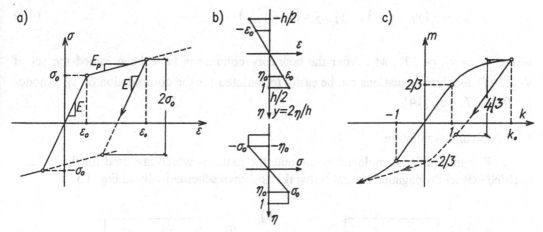

Fig.4.2: a) Stress-strain relationship, b) Strain and stress distribution along beam thickness, c) Bending moment - curvature relationship $m\,(k)$

The cross-sectional bending moment and distribution of strains are defined by the classic formulae:

$$M = \iint_A \sigma y \, dA = \frac{bh^2}{4}\sigma_0 \int_{-1}^{1} s\eta \, d\eta, \qquad \varepsilon = \kappa y = \varepsilon_0 k\eta \;, \tag{4.3}$$

where: $A = b \cdot h$ – area of the beam rectangular cross-section and other dimensionless variables are:

$$m = \frac{4M}{bh\sigma_0^2}, \quad k = \frac{\kappa h}{2\varepsilon_0}, \quad s = \frac{\sigma}{\sigma_0}, \quad \eta = \frac{2y}{h}, \quad \chi = \frac{E_p}{E}. \tag{4.4}$$

After integration of $(4.3)_1$ and introduction of dimensionless variables (4.4) the following formula was derived, cf. Fig. 4.2c:

$$
m = \begin{cases} \dfrac{2}{3}k, & \text{for } |k| \le 1, \\[2mm] \dfrac{2}{3}\chi k + (1-\chi)\left(1 - \dfrac{1}{3k^2}\right)\operatorname{sgn}k, & \text{for } |k| > 1. \end{cases}
\tag{4.5}
$$

The finite difference scheme with a midpoint, cf. Fig. 4.1c, corresponds to the following formulae:

$$
y_{l-1/2} \approx \frac{1}{2}(y_l + y_{l-1}), \quad y'_{l-1/2} \approx (y_l - y_{l-1})/\Delta x_l ,
\tag{4.6}
$$

where: $y_l = v_l, \varphi_l, T_l, M_l$. After the boundary conditions have been joined the set of $4 \cdot (L+1)$ algebraic equations can be easily formulated for the computation of y_l at nodes $l = 0,1,\ldots,L$ – cf. [14].

4.1.2.2. Neural simulation

Formula (4.5) is employed to compute the patterns which are used for the training of BPNN (BackPropagation Neural Network) as shown schematically in Fig. 4.3.

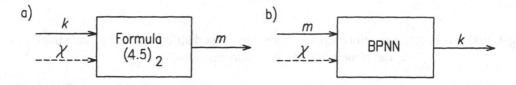

Fig.4.3: a) Application of formula (4.5)₂ to compute patterns for BPNN shown in b)

After a number of numerical experiments the BPNN architecture 1-10-1 was adopted with the binary sigmoid for all the neurons and for the fixed values of the strain-hardening parameter $\chi = 0.0, 0.1$ (i.e. different BPNNs were trained for $\chi = 0.0$ and $\chi = 0.1$, respectively). In order to cover the range $k \in [1.0, 20.0]$ four BPNNs were assumed for each value of χ and the number of training patterns $L=288$. The input and output values were scaled to the range $[0.1, 0.9]$ and then the Rprop learning method (cf. formulae (1.36-37)) was used in the frame of SNNS computer system [13]. The training was limited to 2000 epochs. The trained networks were tested on about 500 patterns and it was proved that the relative error for predicted k was below 1.0%, cf.[5].

4.1.2.3. Application of hybrid NN/FD program

In order to examine the implemented neural-network/finite-difference hybrid program a cantilever beam under a concentrated load P was discussed in [7]. The beam was of constant cross-section with $b*h = 0.012\,m * 0.025\,m$ and length $le = 1.0\,m$. The material characteristics were: $\sigma_o = 210\,MPa$, $E = 2.1 \cdot 10^5\,MPa$. The distance of finite different stations was variable. At the clamped edge for $\xi = x/le \in [0.0\ , 0.5]$ the distance $\Delta\xi_l = \Delta x_l / le = 0.005$, then for $\xi \in [0.05, 0.4]$ the corresponding distance was $\Delta\xi_l = 0.05$ and $\Delta\xi_l = 0.1$ for $\xi \in [0.4, 1.0]$.

The first example is associated with a force increasing monotonically. The elastic load carrying capacity is $P_{el} = bh^2\sigma_o / (6le) = 262.5\,N$. The corresponding displacement at the load application is $v_1^{el} = v(\xi = 1; P_{el}) = 2le^2\sigma_o / (3Eh) = 0.0267\,m$.

In case of elastic perfect plastic material, i.e. for $\chi = 0$, the relation (4.5)$_2$ can be inverted into the following form:

$$k = \frac{1}{\sqrt{3(1-|m|)}}\,\text{sgn}\ m\ .\qquad\qquad(4.7)$$

The corresponding differential equation $v'' = -k(m)$ can be analytically integrated (cf. [10]) and give the following displacement $v_1^{pl} = 20v_1^{el}/9 = 0.0593\,m$, associated with the plastic load carrying capacity $P_{pl} = 1.5\,P_{el} = 393.7\,N$.

In Fig. 4.4 the diagrams of equilibrium paths $v_1(P;\chi)$ are shown for $\chi = 0.0\ , 0.1, 1.0$ computed by the computational program (called as FD, i.e. Finite Difference program) and the hybrid BPNN/FD program (called as FD-H).

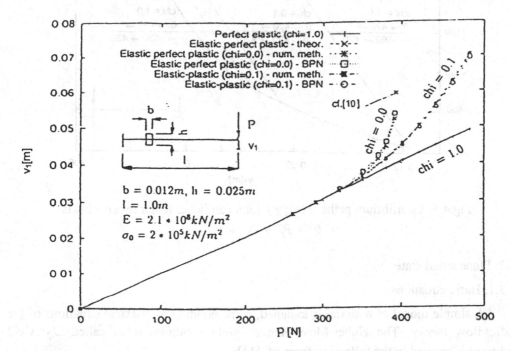

Fig.4.4: Equilibrium paths $v_1(P;\chi)$ for an elastoplastic cantilever under concentrated load

In case of $\chi = 1.0$, i.e. for linear elastic material, only formula (4.5)$_1$ is active and there are, of course, no differences between the results obtained by FD or FD-H.

The case $\chi = 0.0$ corresponds to elastic perfect plastic material and the results obtained by FD and FD-H are very close to each other. The last point at the curve $v_1(P)$ was computed for $P = 392.0 \, N$ associated with $k_1 \approx 20.0$. The point corresponding to the plastic load carrying capacity P_{pl} can be reached at $k \to \infty$.

The FD and FD-H programs were also used to analize the same cantilever under reverse loads. The results correspond to the loading paths $P(v_1 ; \chi)$ for $\chi = 0.1 ; 1.0$ and the following load programs: $0 \to P_1 \to P_2 \to P_1$, where: $P_1 = 420.0 \, kN$, $P_2 = -420.0$, $-560.0 \, kN$.

In Fig. 4.5 closed equilibrium paths are shown. The hybrid simulation by the program FD-H gives practically the same results as the finite program FD gives.

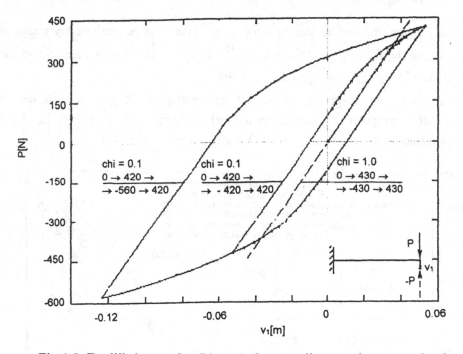

Fig.4.5: Equilibrium paths $P(v_1 ; \chi)$ for a cantilever under reverse loads
$$0 \to P_1 \to P_2 \to P_1$$

4.1.3. Plane stress state

4.1.3.1. Basic equations

A simple model of material is assumed, corresponding to classical equations of the plastic flow theory. The Huber-Mises-Hencky yield condition (also called J_2 yield condition) is applied in the following form, cf. [11]:

$$F \equiv \frac{1}{2}\boldsymbol{\sigma}^T \boldsymbol{P}\boldsymbol{\sigma} - \frac{1}{3}\sigma_e^2(\varepsilon_p) = 0 \ , \tag{4.8}$$

where:

$$\boldsymbol{\sigma}_{(3\times1)} = (\sigma_x, \sigma_y, \tau_{xy}), \quad \boldsymbol{P}_{(3\times3)} = \begin{bmatrix} 2/3 & -1/3 & 0 \\ -1/3 & 2/3 & 0 \\ 0 & 0 & 2 \end{bmatrix}, \quad \sigma_e = \sigma_o + H\varepsilon_p \ . \tag{4.9}$$

In $(4.9)_3$ there are: H – linear strain-hardening parameter, ε_p – Odqvist's parameter cumulated of increments:

$$\Delta\varepsilon_p = \Delta\lambda\sqrt{\frac{2}{3}\boldsymbol{\sigma}^T \boldsymbol{P}\boldsymbol{\sigma}} \ , \tag{4.10}$$

where: $\Delta\lambda$ – increment of plastic multiplier λ .

On the base of assumption of small strains the vector of strain increment $\Delta\boldsymbol{\varepsilon}$ can be split into elastic and plastic parts $\Delta\boldsymbol{\varepsilon}^e$ and $\Delta\boldsymbol{\varepsilon}^P$:

$$\Delta\boldsymbol{\varepsilon} \equiv \left\{\Delta\varepsilon_x, \Delta\varepsilon_y, \Delta\gamma_{xy}\right\} = \Delta\boldsymbol{\varepsilon}^e + \Delta\boldsymbol{\varepsilon}^P \ , \tag{4.11}$$

where $\Delta\boldsymbol{\varepsilon}^e$ obeys Hooke's law and $\Delta\boldsymbol{\varepsilon}^P$ can be computed according to the associated flow rule:

$$\Delta\boldsymbol{\varepsilon}^e = \boldsymbol{E}^{-1}\Delta\boldsymbol{\sigma} \ , \quad \Delta\boldsymbol{\varepsilon}^P = \Delta\lambda\frac{\partial F}{\partial\boldsymbol{\sigma}} \equiv \Delta\lambda\,\boldsymbol{P}\,\boldsymbol{\sigma} \ . \tag{4.12}$$

The elasticity matrix $\boldsymbol{E}^{-1}$ corresponds to the initially isotropic material with the local stiffness matrix $\boldsymbol{E}$:

$$\boldsymbol{E} = \frac{E}{1-v^2}\begin{bmatrix} 1 & v & 0 \\ v & 1 & 0 \\ 0 & 0 & (1-v)/2 \end{bmatrix} , \tag{4.13}$$

where: E – Young's modulus, v – Poisson's coefficient.

The main problem of the analysis at the point level (Gauss' points in plane finite elements) is how to compute the current stress vector $\boldsymbol{\sigma}^D$ and consistent modular matrix E_B^{ep} for the known vectors of stresses $\boldsymbol{\sigma}^A$ and strain increment $\Delta\boldsymbol{\varepsilon}$. In Fig. 4.6 the so-called Return Mapping Algorithm (RMA) is shown, cf. [12]. It depends on the elastic prediction of the stress vector $\boldsymbol{\sigma}^* = \boldsymbol{\sigma}^A + E\Delta\boldsymbol{\varepsilon}$ and orthogonal projection on the current yield surface F^D :

$$F^D \equiv F\left(\boldsymbol{\sigma}^D, \Delta\lambda_D\right) = 0 \ . \tag{4.14}$$

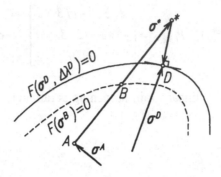

Fig.4.6: Elastic prediction and orthogonal projection on current yield surface in RMA
(Return Mapping Algorithm)

The main problem of RMA is the analysis of highly nonlinear equation (4.14) in order to compute the increment of yielding parameter $\Delta\lambda_D$ (the stress vector $\boldsymbol{\sigma}^D$ is computed simultaneously). It enables us to calculate subsequent variables:

$$\Delta\lambda_D \rightarrow I^D = \left(I + \Delta\lambda_D EP\right)^{-1} \ ,$$

$$\boldsymbol{\sigma}^D = I^D \boldsymbol{\sigma}^* \ ,$$

$$\Delta\varepsilon_p^D = \Delta\lambda_D \left(\frac{2}{3}\left(\boldsymbol{\sigma}^D\right)^T P\boldsymbol{\sigma}^D\right)^{1/2} \ , \quad \sigma_e^D = \sigma_e^B + H\Delta\varepsilon_p^D \tag{4.15}$$

$$E^D = I^D E \ , \quad s = P\boldsymbol{\sigma}^D \ , \quad a = E^D s \ , \quad A = \frac{4}{9}\left(\sigma_e^D\right)^2 H \Big/ \left(1 - \frac{2}{3}H\Delta\lambda_D\right) \ ,$$

$$E_D^{ep} = E^D - a a^T \Big/ \left(A + s^T a\right).$$

4.1.3.2. Neural procedure for RMA

BPNN was used for simulation of RMA (Return Mapping Algorithm). The input and output vectors

$$x_{(7\times1)} = \left\{\overline{\boldsymbol{\sigma}}^B, \ \Delta\overline{\boldsymbol{\varepsilon}} \ ; \ \chi\right\} \ , \quad y_{(4\times1)} = \left\{\Delta\overline{\lambda} \ , \ \overline{\boldsymbol{\sigma}}^D\right\}, \tag{4.16}$$

are composed of the following dimensionless variables:

$$\overline{\boldsymbol{\sigma}} = \boldsymbol{\sigma} / \sigma_e^B \ , \quad \Delta\overline{\boldsymbol{\varepsilon}} = E\Delta\varepsilon / \sigma_e^B \ , \quad \Delta\overline{\lambda} = E\Delta\lambda \ , \quad \chi = H / E \ . \tag{4.17}$$

In Fig.4.7 the scheme of BPNN procedure is shown. The strain-hardening parameter χ was fixed. Thus, in fact only six components of the input vector are used. Looking at formulae (4.15), it is clear that the output stress vector $\bar{\sigma}^D$ is a function of the output variable $\Delta\bar{\lambda}$. Despite the correlation of output variables, the output vector y was composed of $\Delta\bar{\lambda}$ and $\bar{\sigma}^D$ in order to have one-to-one correspondence in the mapping $x \rightarrow y$.

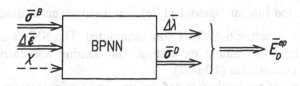

Fig.4.7: Scheme of BPNN procedure

After the computation of $\Delta\bar{\lambda}$ and $\bar{\sigma}^D$ by means of BPNN procedure the consistent stiffness matrix $E_D^{ep} = E \cdot \bar{E}_D^{ep}$ is computed using formulae (4.15).

The patterns for BPNN training and testing were prepared numerically using only constitutive relationships discussed in Point 4.1.3.1. Due to·dimensionless variables (4.17) the computed patterns are independent of current instant. The points corresponding to the stress vector $\bar{\sigma}^B$ are uniformly distributed at the scaled yield surface $\bar{F}^B = 0$. In Fig.4.8 the contour lines of the yield surface are shown for the shear stresses $\pm\bar{\tau}_{xy}^B = const$. For the whole yield surface (for both positive and negative stresses $\bar{\tau}_{xy}^B$) 360 points $\bar{\sigma}^B$ were

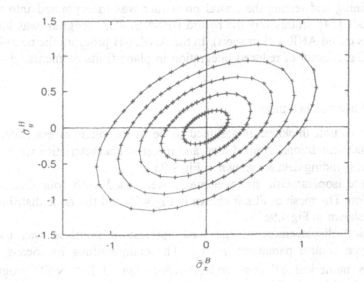

Fig.4.8: Contour lines of the yield surface $\bar{F}^B = 0$ at $\pm\bar{\tau}_{xy}^B = const$. and fixed points $\bar{\sigma}^B$

Fig.4.8: Contour lines of the yield surface $\overline{F}^B = 0$ at $\pm \overline{\tau}_{xy}^B = const.$ and fixed points $\overline{\sigma}^B$ selected. The components of strain vector increment $\Delta \overline{\varepsilon}$ were randomly selected from the range $\Delta \overline{\varepsilon}_x, \Delta \overline{\varepsilon}_y, \Delta \overline{\gamma}_{xy} \in [-10.9, 10.0]$. In this way about $1*10^6$ patterns were formulated and values of $\Delta \overline{\lambda}$ and $\overline{\sigma}^D$ were computed by the use of the return mapping algorithm.

Because of positive and negative values of the output stress vector components $\{\overline{\sigma}_x, \overline{\sigma}_y, \overline{\tau}_{xy}\}^D$ the bipolar sigmoid (1.7e) was used as an activation function. Three-layer BPNN of the structure 6-40-20-4 was formulated. The SNNS computer system [13] was used for the network training and testing. The training was carried out by the Rprop learning method, cf. formulae (1.36-37).

The numbers of randomly selected training and testing patterns were $L = 3349$ and $T = 9044$, respectively. All the output patterns were scaled to the range $[-0.9, 0.9]$ in order to work in good saturation of the activation function. The learning process was continued up to $S = 34000$ epochs. The corresponding Mean-Square-Errors (1.41) were $MSEL \approx 8 \cdot 10^{-5}$ and $MSET \approx 10 \cdot 10^{-5}$, respectively for the network training and testing.

In order to compare the computational time used by the trained neural procedure the same number of 10000 of the same patterns were analyzed by the numerical procedure and neural procedure. The performed computations pointed out the fact that the neural procedure needed about 40-50% of the time consumed by the numerical procedure. The larger the value of the components of strain incremental vector $\Delta \varepsilon_i$ the more iterations the numerical procedure needs, whereas the computational time of neural procedure is practically constant (independent of the values of $\Delta \varepsilon_i$).

After training and testing the neural procedure was incorporated into the ANKA finite element code [14]. In this way the hybrid BPNN/ANKA program was implemented (the program was called ANKA-H in short). In the ANKA-H program the neural procedure is used at each Gauss point of reduced integration in plane finite elements taken from the ANKA library.

4.1.3.3. Analysis of a notched plate

The plate of unit thickness is assumed to be in the plane stress state. The plate geometry, boundary conditions, applied load and material characteristics were taken from [15]. All the corresponding data are shown in Fig.4.9a.

The 8-node isoparametric finite elements were used with four Gauss points of reduced integration. The mesh of FEs is shown in Fig.4.9b and the node distribution in the notch vicinity is shown in Fig.4.9c

The vertical displacement of the load application point A was used as the displacement type control parameter $\tau = v_A$. The computations by means of ANKA programs (purely numerical FE code and ANKA-H hybrid BPNN/FE program) were performed at the fixed length of step $\Delta v_A = -0.02 \, mm$ for $v_A \in [-1.0, 0.0] \, mm$.

displacements v_A and v_{294} (at the notch root), respectively. The neural procedure was trained at the strain hardening parameter $\chi \in 0.032$ but its generalization properties are quite satisfactory for other values $\chi \in [0.0 , 0.07]$. For parameter $\chi = 0.10$ there are visible discrepancies for the load paths computed by the FE code ANKA and ANKA-Ḣ hybrid program. In the case of elastic perfect plastic material, i.e. for $\chi = 0$, the notch root (node 294 in Fig.4.9c) is practically immovable.

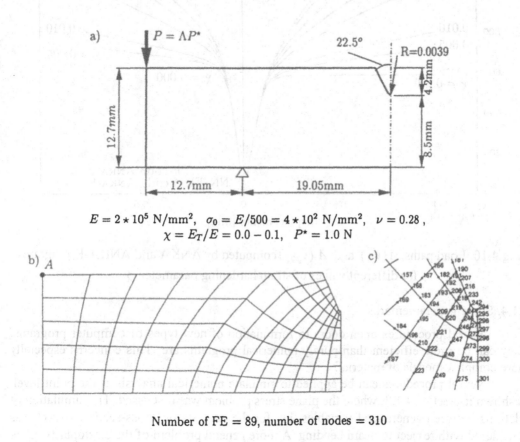

$$E = 2 \star 10^5 \text{ N/mm}^2, \quad \sigma_0 = E/500 = 4 \star 10^2 \text{ N/mm}^2, \quad \nu = 0.28 ,$$
$$\chi = E_T/E = 0.0 - 0.1, \quad P^* = 1.0 \text{ N}$$

Number of FE = 89, number of nodes = 310

Fig.4.9: a) Geometry, loads and material characteristics, b) FE mesh of a half of plate, c) Nodes in vicinity of notch

In order to compare processor time used by programs the computation was carried out for the strain-hardening parameter $\chi = 0.005$ and 65 steps $\Delta v_A = -0.02$ mm. In case of ANKA the processor time was 74 sec at 145 global iterations. The processor time used by the ANKA-H program was 38 sec at 83 iterations. Then the computation was repeated for $\chi = 0.07$. The processor time was 78 sec. at 146 iterations for ANKA and the corresponding figures were 64 sec. at 117 iterations if ANKA-H program was explored.

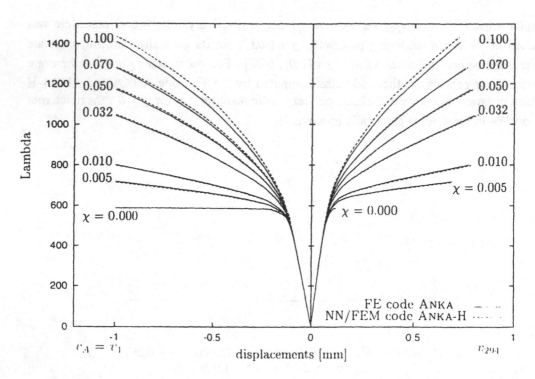

Fig.4.10. Load paths $\Lambda(v_A)$ and $\Lambda(v_{294})$ computed by ANKA and ANKA-H programs
for different values of strain hardening parameter χ

4.1.4. Some general remarks

Neural procedures open doors to formulation of new types of computer programs. They can be more efficient than purely numerical programs are. This concerns especially more complex models of material.

Neural procedures can be applied to simulate numerical analysis at the point level, as shown in Section 4.1.3, where the plane stress problem was discussed. The simulation of relations between generalized variables, i.e. for the analysis on cross-sectional level was considered with respect to beam bending. A more general problem of the elastoplastic plate bending was discussed in [9] where a neural procedure was introduced to simulate the computation of cross-section stiffnesses.

One of the main problems is the formulation of training and testing patterns in order to design neural procedures of good generalization properties.

Besides BPNNs other types of neural networks can be efficiently used in hybrid systems. This problem is discussed in next Sections where the Hopfield neural network is modified in order to analyze unilateral constraints. The parameters of such a network are taken from the standard FE program.

4.2. NEURAL NETWORK APPROACH TO THE ANALYSIS OF QP PROBLEMS

The Hopfield neural networks can be directly used to the analysis of optimization problems. The NN approach consists in the computation of state variables $x_i = f_i(u_i)$ such that local minimum of the network energy E, i.e. a stable state of the network, is identical to the solution of the analyzed minimum problem.

In what follows Eq. (1.81) is used with determined parameters τ_i, w_{ij}, q_i [*)]

The stable state is achieved by integrating the differential equations

$$\tau_i \frac{du_i}{dt} = -u_i(t) + \sum_{j=1}^{N} w_{ij} x_j(t) + q_i \ , \tag{4.18}$$

$$x_i(t) = f_i(u(t)) \qquad \text{for } i = 1,...,N. \tag{4.19}$$

The corresponding energy function is

$$E(t) = -\frac{1}{2} \sum_{i=1}^{N} \sum_{j=1}^{N} w_{ij} x_i x_j + \sum_{i=1}^{N} \int_0^{x_i} f_i^{-1}(\xi) d\xi - \sum_{i=1}^{N} q_i x_i \ . \tag{4.20}$$

In general function (4.19) is nonlinear and numerical methods have to be applied (e.g. Runge-Kutta method) for integrating Eqs (4.18). Starting from initial values $x_i(0)$ the integration process is continued with steps Δt_m up to value $t = T$ corresponding to $(dE / dt)_{t=T} \approx 0$.

The Hopfield type neural networks have been successfully applied to the analysis of many optimization problems starting from paper by Hopfield and Tank [16].

In this Chapter the attention is focused on the analysis of two *Quadratic Programming Problems* (QPPs):

i) *Unconstrained problem* (bilateral problem): Find $x \in \mathcal{R}^n$ such as to solve the problem

$$\min \left\{ \frac{1}{2} x^T M x - q^T x \right\} , \tag{4.21}$$

ii) *Constrained problem* (unilateral problem): Find $x \in \mathcal{R}^n$ such as to solve the problem

---

[*)] In papers by Panagiotopoulos et al., quoted in References, notations of quantities correspond to the circuit analog network, i.e. $x_i = V_i$, $w_{ij} = T_{ij} R_i$, $q_i = I_i R_i$, $\tau_i = C_i R_i$ and then $C_i = R_i = 1$ was assumed.

$$\min\left\{\frac{1}{2}x^T M x - q^T x \mid x \geq 0\right\}.$$ (4.22)

Here $M = \{\mu_{ij}\}$ is a given matrix and $q = \{q_i\}$ is a given vector.

Now we focus our attention on the formulation of fictitious neural networks the stable state of which corresponds to the solutions of (4.21) or (4.22). The discussion below is based on papers by Panagiotopoulos et al. [17-19].

In case (i) of bilateral QPP the following substitution can be made:[*]

$$w_{ij} = \begin{cases} -\mu_{ij} & \text{for } i \neq j, \\ -\mu_{ij} + 1 & \text{for } i = j, \end{cases}$$ (4.23)

$$x_i = f_i(u_i) = u_i.$$ (4.24)

Further, $\tau_i = 1$ can be assumed for the sake of simplicity. From (4.18) and (4.24) the following evolution equation results:

$$\frac{dx_i}{dt} = \sum_j w_{ij} x_j - x_i + q_i.$$ (4.25)

At the stable state it is $dx_i/dt = 0$ and after substituting (4.23) into (4.25) we have

$$0 = -\sum_{i \neq j} \mu_{ij} x_j - \mu_{ii} x_i + x_i - x_i + q_i.$$ (4.26)

Writing (4.26) in the matrix form it is evident that the stable state corresponds to the linear equation

$$M x = q,$$ (4.27)

which yields the solution of the minimum problem (4.21).

In order to consider case (ii), related to the unilateral QPP, the following activation function is assumed:

$$x_i = f_i(u_i) = \begin{cases} u_i & \text{for } u_i > 0, \\ 0 & \text{for } u_i \leq 0, \end{cases}$$ (4.28)

---

[*] In comparison with assumption $(1.68)_1$ the nonzero diagonal synaptic weights $w_{ii} \neq 0$ are introduced in $(4.23)_2$.

instead of the identity function (4.24).

Let us assume the function of energy $E(t)$ for $t > 0$ with the subsidiary condition $x_i(t) \geq 0$. The discussion below corresponds to the stable state related to the solution

$$\text{stat}\left\{ E(t) \mid x_i(t) \geq 0, \ i = 1,\ldots,N \right\}. \tag{4.29}$$

First the Lagrangian $L(t)$ is formulated:

$$L(t) = E(t) - \sum_i \lambda_i \, x_i(t). \tag{4.30}$$

The equivalent conditions for stationarity problem (4.29) are:

$$\frac{dE}{dt} - \sum_i \lambda_i \frac{dx_i}{dt} = 0, \tag{4.31}$$

$$V_i \geq 0, \ \lambda_i \geq 0, \ \lambda_i x_i = 0. \tag{4.32}$$

Now some manipulations are made, taking into account (4.20) and (4.18):

$$\frac{dE}{dt} = \sum_i \frac{\partial E}{\partial x_i} \frac{dx_i}{dt} = \sum_i \left(-\sum_j w_{ij} x_j + u_i - q_i \right) \frac{\partial x_i}{\partial u_i} \frac{du_i}{dt} = -\sum_i \tau_i \left(\frac{du_i}{dt} \right)^2 \theta_i, \tag{4.33}$$

where:

$$\theta_i = \frac{dx_i}{du_i} = \begin{cases} 1 & \text{for} \quad u_i > 0, \\ 0 & \text{for} \quad u_i \leq 0. \end{cases} \tag{4.34}$$

Thus, $dE/dt \leq 0$ for every $t > 0$ but from (4.31) we have:

$$\frac{dE}{dt} = \sum_i \lambda_i \frac{\partial x_i}{\partial u_i} \frac{du_i}{dt} = \sum_i \lambda_i \theta_i \frac{du_i}{dt}. \tag{4.35}$$

For those i for which $u_i = x_i > 0$ at time t we have $\theta_i = 1$ and $\lambda_i = 0$, according to (4.32)₃. At the same time t for other i we have $\theta_i = 0$, so all terms in (4.35) disappear and, thus

$$\frac{dE}{dt} = 0. \tag{4.36}$$

Eqs (4.33) and (4.36) imply that for every i either $u_i \leq 0$, i.e. $x_i = 0$ or if $u_i > 0$ we have $u_i = x_i$ and $du_i / dt = dx_i / dt = 0$, i.e. $x_i = const$. Thus, Hopfield's results can be extended for the case of inequality constraints.

Let us now return to the stationarity condition (4.31) written in the form:

$$\sum_i \left(\frac{\partial E}{\partial x_i} - \lambda_i \right) \frac{dx_i}{dt} = 0 \quad \text{for} \quad \forall t \quad \text{and} \quad i = 1,\ldots,N , \tag{4.37}$$

and let us consider the minimum problem

$$\min \left\{ E(x_i,\ldots,x_N) \mid x_i \geq 0, i = 1,\ldots,N \right\} . \tag{4.38}$$

Since for every t and for all i there is $dx_i / dt = 0$, the solution of (4.38) implies

$$\frac{\partial E}{\partial x_i} = \lambda_i , \quad \lambda_i \geq 0, \quad x_i \geq 0, \quad \lambda_i x_i = 0 . \tag{4.39}$$

Thus, the solution of (4.38) satisfies the stationarity problem (4.29) and through the substitutions of (4.23) the constrained problem (4.22) is also satisfied.

The Hopfield-Tank approach above mentioned was extended for the case of inequality constraints. Due to this approach the treatment of constrained minimum problems was put on a sound basis related to the formulation of constrained QP problem (4.22) and then to the analysis of QPP in the neural environment.

In what follows the formulation of mechanical problems as constrained QP problems of form (4.22) and then their analysis in neural environment is called the *Panagiotopoulos approach*.

4.3. PANAGIOTOPOULOS APPROACH TO THE ANALYSIS OF BILATERAL AND UNILATERAL ELASTIC PROBLEMS

4.3.1. Unilateral constraints in elastic solids

Let us consider an elastic body Ω with boundary Γ composed of three parts Γ_U, Γ_F and Γ_S, where: Γ_U – boundary with displacement conditions, Γ_F – boundary with force (strain) conditions, Γ_S – boundary with inequality conditions, cf. Fig.4.11a. Different boundary and interface conditions are discussed in [18,20].

In case of rigid support and unilateral constraints at the boundary Γ_S the inequality conditions (Signorini-Fichera boundary conditions) are as shown in Fig.4.11d:

$$\begin{aligned}
&\text{if} \quad u_N < 0 \quad \text{then} \quad S_N = 0 , \\
&\text{if} \quad u_N = 0 \quad \text{then} \quad S_N < 0.
\end{aligned} \tag{4.40}$$

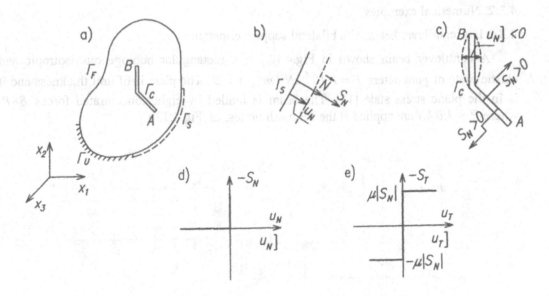

Fig.4.11: a) Solid body, b) Normal traction S_N and normal displacement u_N
at surface Γ_S, c) Positive directions at u_N] and S_N at crack interface Γ_C,
d) Unilateral contact law, e) Friction law.

The friction boundary conditions are, cf. Fig.4.11e:

$$
\begin{aligned}
&\text{if} \quad |S_T| < \mu |S_N| \quad &&\text{then} \quad u_T = 0, \\
&\text{if} \quad |S_T| = \mu |S_N| \quad &&\text{then} \quad u_T > 0, \\
&\text{if} \quad |S_T| = -\mu |S_N| \quad &&\text{then} \quad u_T < 0.
\end{aligned}
\qquad (4.41)
$$

where: μ – Coloumb's coefficient of friction.

The interface detachment problem can be related to the relative normal displacement u_N] and normal traction S_N shown in Fig.4.11c:

$$
\text{if} \quad u_N] \le 0 \quad \text{then} \quad S_N = 0.
\qquad (4.42)
$$

The friction problem includes the constraint

$$
\text{if} \quad |S_T| \le \mu |S_N| \quad \text{then} \quad u_T] = 0.
\qquad (4.43)
$$

4.3.2. Numerical examples

4.3.2.1. A cantilever beam with bilateral support constraints

A cantilever beam shown in Fig.4.12a is a rectangular homogenous isotropic and elastic plate of parameters $E = 1 \cdot 10^4 \ kN \ / \ m^2$, $v = 0$. The plate is of unit thickness and it is in the plane stress state [17]. The beam is loaded by eight concentrated forces $8 \times P$, where $P = 1.0 \ kN$ are applied at the FE mesh nodes, cf. Fig.4.12b.

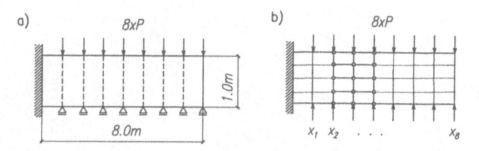

Fig.4.12: a) A cantilever beam, b) FE mesh and state variables x_i

In the example 32 four-node isoparametric, plane stress elements with 2x2 Gauss points were used. Matrix M in (4.27) is the influence matrix of unknown reactions x_i for $i = 1,...,8$. The components of vector q in (4.27) correspond to displacements of the reaction points when the supports are removed, due to the external loads.

The resulting set of compatibility equations was solved by the Gauss-Seidel method and the values of reactions (also called a solution of the problem) are in kN: $x_1 = 7.430$, $x_2 = 10.25$, $x_3 = 10.17$, $x_4 = 10.01$, $x_5 = 9.811$, $x_6 = 9.396$, $x_7 = 11.73$, $x_8 = 9.055$. They are associated with a minimum of the complementary energy $E_c = -10.8478 kNm$.

The neural network approach corresponds to the integration of a set of differential equations (4.25) and (4.24). The equations were solved numerically by the Runge-Kutta fourth-order method with step $\Delta t = 0.5$. Applying the initial values, even very close to the solution, the convergence was very slow in this example. After 3500 steps the minimum value of complementary energy was -10.8476 but the values of reactions were: $x_1 = 7.694$, $x_2 = 10.634$, $x_3 = 10.006$, $x_4 = 10.009$, $x_5 = 10.066$, $x_6 = 8.904$, $x_7 = 11.894$, $x_8 = 9.111$. It is worth noting that this energy minimum value was obtained first at a step between steps 2000-2100 and after step 2100 the minimum value remained in all steps the same.

The same example solved on a computer with no RISC architecture (a classical 386-computer) needed steps three times as much. Note that the neural network approach is not influenced by the round-off errors of the computation as it obvious from its iterative character.

4.3.2.2. A crack with classical interface conditions

The structure shown in Fig. 4.13a is a plane elastic, homogoneous and isotropic body of material parameters: $E = 2.1 \cdot 10^5\,kN\,/\,m^2$, $v = 0.24$ and thickness $t = 16$ cm. The structure is loaded by four pairs of forces P which tend to open a crack of a given initial length [18].

The structure was discretized by four-node isoparametric, linear interpolation elements and by triangle constant stress elements, cf. Fig.4.13b. Around the crack-tip 12-node isoparametric, cube interpolation elements were used which collapsed into triangular elements after an appropriate mode coupling [21].

The classical FE displacement method was used to compute components μ_{ij} and q_i of the matrix M and vector q of the linear set of Eqs (4.26). The solution of the problem was obtained by the Gauss-Seidel method at the minimum value of potential energy $E_p = -1343.3kNm$. The corresponding displacement and stress fields are shown in Figs 4.13b,c.

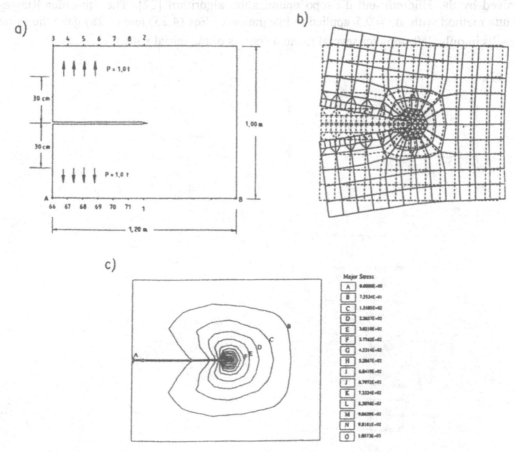

Fig.4.13: a) Geometry and applied loads, b) FE mesh and displacement field,
c) Field of major stresses.

As in the previous example the Runge-Kutta fourth-order method with step $\Delta t = 0.5$ was used. The iteration process was quite slow. The results close to the solution obtained by the Gauss-Siedel method (with insignificant differences) were computed after 5500 steps. The minimum value of potential energy $-1343.3\ kN$ was reached between 2300 and 3000 steps and then the values of displacements were improved at unchanged value of potential energy.

4.3.2.3. A crack with detachment

In this example the structure from the previous example is considered but different loads are applied, cf. Fig.4.14a. At each node of the crack relations (4.39) hold together with the constraints $u_T] = 0$ in the tangential direction.

After normalization and elimination of all the bilateral degrees of freedom an inequality constrained minimum problem was formulated [18]. The complementary energy was expressed with respect to the normal forces of 18 pairs of the contact nodes of the crack interface, cf. Fig.4.14b. The normalized optimization problem of type (4.22) was solved by the Hildreth and d'Esopo optimization algorithm [22]. The 4th-order Runge-Kutta method with $\Delta t = 0.5$ applied to integration of Eqs (4.25) and (4.28) gave the same results in only 250 steps for several random choices of the initial values.

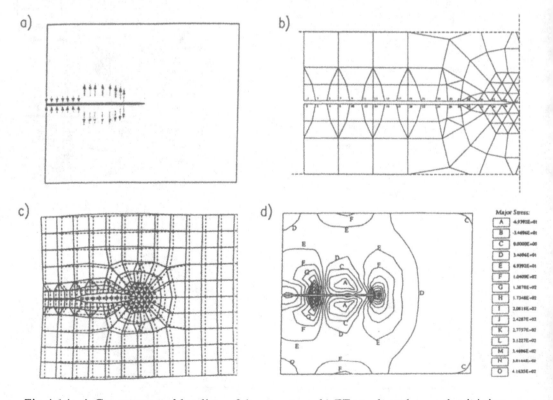

Fig.4.14: a) Geometry and loading of the structure, b) FE mesh at the crack vicinity, c,d) Displacement and major stress fields.

It is worth noting that for this example which is more complicated than the previous one 'in the classical sense', due to inequality subsidiary conditions, the neural network approach is very efficient and shows very good convergence. The inequality conditions do not complicate the numerical treatment of the problem; on the contrary – the activation function (4.28) reduces the search to minimum.

In Figs.4.14c,d the corresponding displacement and major stress fields are depicted.

4.3.2.4. A unilateral 3D contact problem

The problem is associated with a small part of a real project concerning the stability study of the Holy Rock of Golgotha in the Saint Sepulchre Church in Jerusalem, cf. Fig.4.15a. A group under P.D. Panagiotopoulos' supervision carried out this project in 1990. The methods of inequality mechanics were used because of the large interfaces of known geometry in the rock mass presenting large detachments and sliding friction effects.

In this example a special problem is discussed in which the interface has only unilateral contact and no friction [17]. Only the weight of the structure is taken into account. The structure was discretized by eight-node isoparametric brick-elements, by six-node prisms and by four-node tetrahedral elements (totally 900 DOF). After normalization and condensation an inequality constrained minimum problem was formulated with respect to the normal forces at 25 pairs of the interface nodes, cf. Fig.4.15b.

The solution was obtained by the classical quadratic optimization methods of Hildreth and d'Esopo [22]. The same results were given applying the 4-th order Runge-Kutta method with $\Delta t = 0.5$ to the integration of Eqs (4.25) and (4.28) just in 40 steps on a classical 386-computer for several random choices of the initial values. The minimum value of energy of the 40-th step was already obtained at the 24-th step, cf. Fig.4.15c. The corresponding values of reactions differed only slightly between the two steps mentioned. On a RISC computer (HP 720) function (4.28) led to 1/3-1/4 computation time compared with the computation time needed for the structure with function (4.24).

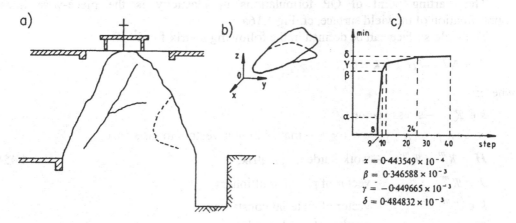

Fig.4.15: a) The form of the Holy Rock of Golgotha in Jerusalem,
b) Schematic form of interface, c) Convergence of the neural network method

4.3.2.5. Conclussions from numerical examples

The numerical examples are a basis to draw the following conclusions [17,18]:

a) The neural approach even with initial values very close to the solution, converges quite slowly in the case of bilateral problems.

b) The quadratic and convex programming algorithms have substantial storage requirements for large matrices; this is not the case in neural network treatment.

c) The numerical analysis of the circuit differential equation (4.18) has better stability properties for large-scale problems than the convex programming algorithms. The neural approach lacks the influence of the round-off errors. These properties give more reliable numerical results especially when the constrained QP problems are computed in the neural environment.

d) The numerical experience shows that singularity does not influence considerably the numerical procedure related to the solution of the initial value problem (4.18). Thus, the neural network approach gives more reliable results around the crack-tip than the classical singular FEM. This conclusion needs a theoretical verification.

4.4. PANAGIOTOPOULOS APPROACH TO THE ANALYSIS OF ELASTOPLASTIC STRUCTURES

The Panagiotopoulos approach was successfully applied to the analysis of elastoplastic structures. In the paper by Avdelas et al. [19] the Hopfield-Tank analog was used for the analysis of a QP problem formulated on the basis of Maier's paper [23]. In what follows this formulation is presented in short and then an example taken from [19] is discussed.

4.4.1. QP problems for elastoplasic structures

The formulation of QP problems for the analysis of elastoplastic solids and structures was mainly developed by Maier in 1970's, cf. references in [24]. The formulation corresponds to both the holonomic (reversible) problems and incremental elastoplasticity.

The starting point of QP formulations in plasticity is the piece-wise linear approximation of the yield surface, cf. Fig.4.16a.

The yield surface can be defined in the following matrix form[*] :

$$F = N^T s - H \lambda - k = 0 , \tag{4.44}$$

where:

$s \in \mathcal{R}^r$ – stress vector,

$N = [N_1, \dots , N_m] \in \mathcal{R}^r \times \mathcal{R}^m$ – matrix of unit vectors in the s space,

$H \in \mathcal{R}^m \times \mathcal{R}^m$ – work-hardening matrix,   (4.45)

$\lambda \in \mathcal{R}^m$ – vector of plastic multipliers,

$k \in \mathcal{R}^m$ – vector of material constants,

m – number of yield modes.

---

[*] Notation from [17-19] is used to make papers by Panagiotopoulos and his associates easier to comprehend.

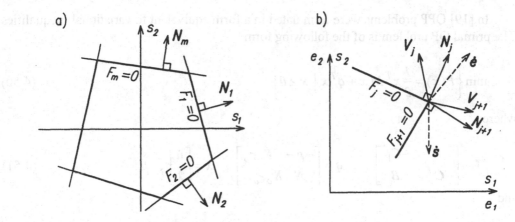

Fig.4.16: a) Piece-wise linear yield surface,
b) Vectors of rates at yield corner

The selection of matrix H and vector k enables us to model the behaviour of different materials and work-hardening rules, cf. [24]. Eq. (4.44) can also be used for generalised variables, e.g. in the case of structural analysis cross-sectional variables can be used.

In the case of elastoplastic analysis, incremental formulation is usually used. The corresponding constitutive equations are of the following form [23,25]:

$$\dot{e} = \dot{e}_o + \dot{e}_E + \dot{e}_P ,$$
$$\dot{e}_E = F_o \dot{s} ,$$
$$\dot{e}_P = V \dot{\lambda} , \tag{4.46}$$
$$\dot{F} = N^T \dot{s} - H \dot{\lambda} ,$$
$$\dot{\lambda} \geq 0, \ \dot{F} \leq 0, \ \dot{F}^T \dot{\lambda} = 0,$$

where $(\dot{\ })$ are rates of vectors used. Besides notation (4.45) there are used in (4.46) the following matrices and vectors:

$$
\begin{aligned}
&e_o, \ e_E, \ e_P && - \text{vectors of initial, elastic and plastic strains, respectively,} \\
&F_o && - \text{flexibility matrix,} && (4.47) \\
&V && - \text{matrix of gradients of plastic potentials, cf. Fig. 4.16b.}
\end{aligned}
$$

Equilibrium and compatibility equations are written under the assumption of small displacements (more general, nonlinear equations can be found in [19,25]):

$$G \dot{s} = \dot{p}, \quad \dot{e} = G^T \dot{u} , \tag{4.48}$$

where:

$$
\begin{aligned}
&G - \text{equilibrium matrix}, \\
&p, u - \text{load and displacement vectors, respectively.} && (4.49)
\end{aligned}
$$

In [19] QPP problems were formulated in a form equivalent to variational inequalities. The primal QP problem is of the following form:

$$\min\left\{ P(x) = \frac{1}{2}x^T M x + q^T x \mid x \geq 0 \right\}, \tag{4.50}$$

where:

$$M = \begin{bmatrix} K & -C_V \\ -C_N^T & B \end{bmatrix}, \quad q = \begin{bmatrix} -p - GK_o\dot{e}_o \\ N^T K_o e_o \end{bmatrix}, \quad x = \begin{bmatrix} \dot{u} \\ \dot{\lambda} \end{bmatrix}, \tag{4.51}$$

and

$$K_o = F_o^{-1}, \quad K = G K_o G^T, \quad B = H + N^T K_o N, \tag{4.52}$$

$$C_V = G K_o V, \quad C_N = G K_o N.$$

Further considerations are made under the assumption that matrix M is symmetric and positive semidefinite (PSD). The sufficient condition for matrix M to be PSD is that matrix H is PSD (non-softening materials). The assumption that matrix M is symmetric is valid if and only if there exists normality ($V = N$) and reciprocity of the yield mode interaction. Other cases correspond to perfect plasticity, i.e. for $H = 0$ and to noninteracting plastic modes, i.e. for $H \equiv diag[H_j]$.

From the other forms of QP formulations [23] let us only mention the one with the rates of plastic multiplayers $\dot{\lambda}$:

$$\min\left\{ R(\dot{\lambda}) = \frac{1}{2}\dot{\lambda}^T D \dot{\lambda} - \left(N^T \dot{s}_E\right)^T \dot{\lambda} \mid \dot{\lambda} \geq 0 \right\}, \tag{4.53}$$

where:

$$D = H - N^T Z V, \quad Z = K_o G^T K^{-1} G K_o - K_o, \tag{4.54}$$

$$\dot{s}_E = K_o G^T K^{-1} \dot{p} + Z \dot{e}_o. \tag{4.55}$$

The symmetry and semipositiveness of matrix D needs additional assumptions with respect to the work-hardening matrix H and to matrix $-N^T Z N$ (for the case $V = N$).

In case of plastic active processes (no local unloading occurs in the structure), the holonomic form of the QP formulation is useful [19]:

$$\min\left\{ R(\lambda) = \frac{1}{2}\lambda^T D \lambda - \left(N^T s_E - k\right)^T \lambda \mid \lambda \geq 0 \right\}, \tag{4.56}$$

where matrix D is defined in $(4.54)_1$ and s_E corresponds to (4.55) if the vectors p and e_o are used instead of the vectors of rates $\dot{p}$ and $\dot{e}_o$. After computation of λ the total stress vectors can be computed

$$s = s_E + s_P,$$

$$s_E = K_o G^T K^{-1} p + Z e_o, \quad s_P = ZV\lambda.$$

$$(4.57)$$

4.4.2. Elastoplastic analysis of a plane frame

A numerical example was taken from [19]. It refers to the elastoplastic analysis of a plane frame, according to the classical assumption of stiffness method, i.e. the only active generalized stress is the bending moment $s = [M]$. In Fig.4.17a the finite element i is shown with inelastic deformations concentrated at the ends L and R, respectively.

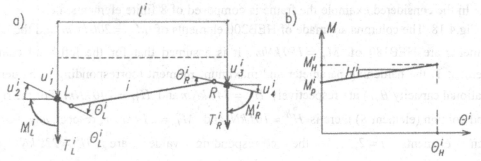

Fig.4.17: a) Frame finite element, b) Plastic properties of the cross-section

The yield condition for the finite element is written in the form

$$F^i_{(4\times 1)} = \left(N^i\right)^T M^i - H^i \lambda^i - k^i = 0.$$

$$(4.58)$$

where:

$$N^i_{(2\times 4)} = \begin{bmatrix} N^i_L & 0 \\ 0 & N^i_R \end{bmatrix}, \quad H^i_{(4\times 4)} = diag[H_i], \quad (k^i)^T = \begin{bmatrix} M^i_p & M^i_p \end{bmatrix}$$

$$(M^i)^T = \begin{bmatrix} M^i_L & M^i_R \end{bmatrix}, \quad (\lambda^i)^T = \begin{bmatrix} \lambda^i_R & \lambda^i_L \end{bmatrix}, \quad e_o = 0,$$

$$(4.59)$$

$$N^i_j = \begin{bmatrix} 1 & -1 \end{bmatrix}, \quad (\lambda^i_j)^T = \begin{bmatrix} \lambda^i_1 & \lambda^i_2 \end{bmatrix}_j \quad \text{for} \quad j = L, R$$

The elastic stiffness matrix K^i_o and compatibility matrix $C^i = \left(G^T\right)^i$ are:

$$K_o^i = \frac{l^i}{6EI^i} \begin{bmatrix} 2 & -1 \\ -1 & 2 \end{bmatrix}, \qquad C^i = \begin{bmatrix} 1/l^i & 1 & -1/l^i & 0 \\ 1/l^i & 0 & -1/l^i & 1 \end{bmatrix}. \tag{4.60}$$

The global matrices and vectors are computed on the basis of corresponding transformations and assemble procedures, cf. e.g. [26]. It is worth emphasising that H is the global diagonal matrix of positive components H^i and for the associative flow rule, i.e. $V = N$, the global matrix $-N^T Z N$ is positive semidefinite. This implies positive definiteness of matrix D, computed according to (4.54)₁.

The QP problem is related to the computation of the vector of plastic multipliers λ, i.e. to the computation of rotations

$$\theta_j^i = (N\lambda)_j^i \qquad \text{for} \quad j = L, R. \tag{4.61}$$

In the considered example the frame is composed of 8 finite elements, i.e. $i = 1,...,8$, cf. Fig.4.18. The columns are made of HEB200 elements of $M_p = 200\,kNm$ and the beam elements are HEB180 of $M_p = 180\,kNm$. It is assumed that for the left-hand column (element 1) the hardening parameter and maximum moment (corresponding to hardening rotational capacity θ_H) are respectively $H^1 = 340\,kNm$ and $H_H^1 = 240\,kNm$. For the right-hand column (element 8) there is $H^8 = 161\,kN$ and $M_H^8 = 219\,kNm$, respectively. For the beam elements $i = 2,...,7$ the corresponding values are $H^i = 229\,kN$ and $M_H^i = 149\,kNm$.

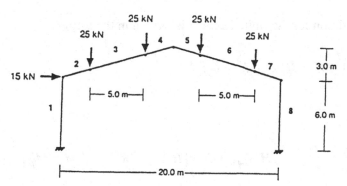

Fig.4.18. Frame geometry, support conditions and applied loads

The computation was performed first by a classical QP algorithm using the Keller direct pivoting method [27] and then the neural approach was applied.

For lower values of the load factor α the frame is in elastic state. This means that $\lambda = 0$ and rotations are blocked, i.e. $\theta = 0$. In Fig.4.19 the values of rotations θ and total moments s at the iL and iR sections are respectively depicted for the load factor α equal

1.402 and *1.583*. The first value $\alpha = 1.402$ is the one for which the frame would collapse if the hardening matrix **H** is zero, i.e. for the elastic perfect plastic material. The second value of load factor $\alpha = 1.583$ corresponds to the first exceeding of the hardening rotational capacity in at least one of the nodes. In the considered case this capacity was nearly reached at the nodes and *8L*, cf. Table 4.1.

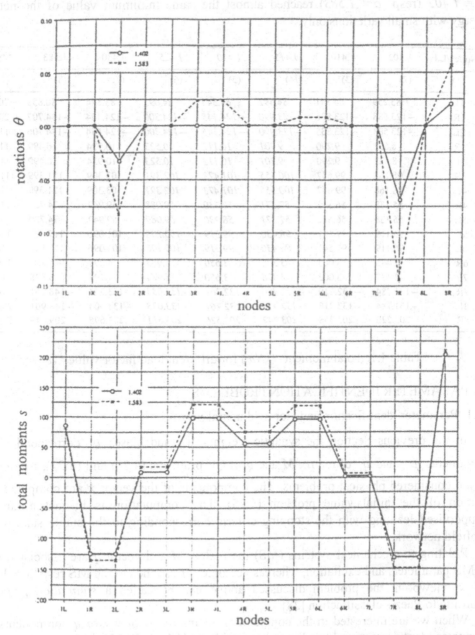

Fig.4.19: a) Rotations - nodes diagram, b) Total moments - nodes diagram

Next the Panagiotopoulos approach was applied to the QP problem (4.56). As in the previous examples Eqs (4.25) were integrated by the Runge-Kutta 4-th order method with step $\Delta t = 0.5$. Several numerical experiments led to the conclusion that the neural approach is very insensitive to changes of the initial conditions. Moreover, the neural algorithm needed 300 steps which after the 30th (resp. the 70th) iteration for the load factor $\alpha = 1.402$ (resp. $\alpha = 1.583$) reached almost the same minimum value of the network energy with small variations only.

Nodes/L.F.	1.402	1.410	1.415	1.420	1.425	1.430	1.583	Mp
(1)	(2)	(3)	(4)	(5)	(6)	(7)	(8)	(9)
1L	85.239	84.841	84.592	84.343	84.093	83.844	80.833	200
1R	−123.588	−123.836	−123.990	−124.145	−124.300	−124.454	−134.707	200
2L	−123.588	−123.835	−123.990	−124.145	−124.300	−124.454	−134.706	116
2R	8.632	9.290	9.701	10.112	10.523	10.934	16.499	116
3L	8.632	9.290	9.701	10.112	10.523	10.934	16.499	116
3R	98.168	99.637	100.555	101.473	102.391	103.309	121.399	116
4L	98.168	99.637	100.555	101.473	102.391	103.309	121.399	116
4R	55.138	56.512	57.371	58.230	59.088	59.947	74.729	116
5L	55.138	56.512	57.371	58.230	59.089	59.947	74.729	116
5R	96.119	97.541	98.430	99.319	100.208	101.096	117.835	116
6L	96.119	97.541	98.430	99.319	100.208	101.096	117.835	116
6R	2.485	3.003	3.326	3.650	3.973	4.297	5.808	116
7L	2.485	3.003	3.326	3.650	3.973	4.297	5.808	116
7R	−131.784	−132.219	−132.490	−132.761	−133.033	−133.304	−148.961	116
8L	−131.784	−132.218	−132.490	−132.761	−133.033	−133.304	−148.961	200
8R	203.223	203.358	203.442	203.526	203.611	203.695	209.049	200

Table 4.1. Total moments (kNm) for different load factor values α

4.5. PARAMETER IDENTIFICATION PROBLEMS

4.5.1. Parameter identification and Hopfield neural network

In the previous Section the synaptic weights w_{ij} and biases q_i correspond to the components μ_{ij} and q_i of matrix M and q in QP problems (4.21) and (4.22), formulated for the considered physical problems. The components of the vector x are computed as a solution of the initial value problem (4.25). The computation is in fact a kind of unsupervised learning with the stopping criteria corresponding to the stable state of the Hopfield network.

Relating to mechanical systems (MS) the matrix M and vector q are associated with the MS parameters and excitation, whereas vector x corresponds to the MS response. From such a viewpoint the problem discussed above can be called a *simulation problem*, according to Paez's classifiaction [28].

When we are interested in the computation of matrix M or vector q components the *parameter identification* problem has to be considered [17-19, 29-32].

The parameter identification problem is the *inverse problem* of computational mechanics. Usually a solution or a part of solution of the direct problem is known and we try to compute those material, load or geometry parameters which give results very close or identical to the prescribed data. Both in the case of bilateral and unilateral problems [32] this problem is formulated as the minimum deviation problem assuming all the relations characterising the physical problem as subsidiary conditions. For instance in case of unilateral constraints, e.g. in crack interfaces with detachment, the identification problem has variational inequalities as state relations.

It is worth emphasizing that in a neural network environment, the parameter identification problem is of a very simple form which can be easily analyzed if an appropriate *supervised learning problem* is formulated. The parameters we want to identify can be related either to matrix M or to vector q in (4.21) or (4.22). Since matrix M is directly related to the synaptic weights w_{ij} a learning rule is introduced in the following form [17]:

$$w_{ij}^{(k+1)} = w_{ij}^{(k)} + \eta^{(k)} r_{ij}^{(k)} , \tag{4.62}$$

where: $\eta^{(k)} > 0$ – adaptive-type learning rate and $r_{ij}^{(k)}$ – reinforcement signal, k – iteration step.

The reinforcement signal can be computed by means of the Widrow-Hoff delta rule:

$$r_{ij}^{(k)} = \left(x_i^* - x_i^{(k)} \right) u_j^{(k)} , \quad u_j^{(k)} = \sum_{j=0}^{N} w_{ij} x_j^{(k)} \quad \text{for} \quad i = 1,\dots, N , \tag{4.63}$$

where: x_i^* – desired response, $w_{io}^{(k)} = 1$, $x_o^{(k)} = q_i^{(k)}$. The solution at the k-th step is considered as an initial value for the computation of the network at the $(k + 1)$ step by means of Eqs (4.25). This procedure of modifying the synapses weights is continued until the desired result $\{x_i^*\}$ is obtained.

4.5.2. Numerical examples

4.5.2.1. Parameter identification problem for a cracked body

For the bilateral crack problem discussed in Point 4.3.2.2. the following parameter identification problem was solved: for which E and ν boundary displacements of the body had certain given values?

In [18] it was assumed that at the crack shown in Fig. 4.13a the horizontal displacements were: $-1.79 \cdot 10^{-5} m$ for point 5-8, $-1.80 \cdot 10^{-5} m$ for points 9-12 and $-1.81 \cdot 10^{-5} m$ for points 13-14, cf. Fig.4.14b. These values were chosen to the corresponding values of the displacements of the example 4.3.2.2 and they constituted the observation vector of the identification problem.

The identification problem, when formulated as a learning problem, gave in less than *100* steps the values $E = 2.49 \cdot 10^5 \, kNm^{-2}$ and $v = 0.28$ for which the displacement and major stress fields were depicted in Figs 4.13b,c.

It is worth noting that if the problem has more data, i.e. if the number of observations increases, this fact does not necessarily mean that the learning algorithm converges in a smaller number of steps, unless the additional observations are identical to the displacements corresponding to the solution of the identification problem.

4.5.2.2. Identification of load factors for an elastoplastic frame

An identification problem is reduced to the computation of values of the load factor α for the frame analized in Point 4.4.2. Let us assume that the given moment distributions are put together in columns (4), (5) and (6) of Table 4.1.

Here overcomplete information is available since the bending moment diagrams of the structure for both the load factors *1.402* and *1.583* are known (indeed the problem could be solved if only one column was given). Thus, the learning rule can be applied in both the forward and backward direction with respect to the load factor α.

In order to avoid lengthy computations the moment distributions for the load factor *1.41* and *1.43* were computed by means of the direct method. Then the iterative procedure (4.63) offers certain corrections of the entries of the QPP matrices which finally yield load factors corresponding to the moment distributions in the columns (4), (5) and (6) of Table 4.1.

The direct solution showed that load factors *1.415, 1.420* and *1.425* obtained through the learning algorithm led to moment distributions which have nearly the same values (up to the third digit) as the moment distributions given in columns (4) to (6).

4.5.3. Parameter identification by interaction of neural networks.

In the majority of identification problems the identification is associated with state vector $s^{(\tau)}$ and control vector $z^{(\tau)}$ where $\tau = 1, \ldots, T$ are time instants.

Let us assume that the *identification problem* is of the following form:

- Find $z^{(\tau)}$, $\tau = 1, \ldots, T$ such that [30]:

$$\sum_{\tau=1}^{T} \left[\sum_{r=1}^{m} \left\| s_r^{*(\tau)} - s_r\left(z^{(\tau)}\right) \right\| \right] \to \min \qquad (4.64)$$

$$M\left(z^{(\tau)}\right)x = q\left(z^{(\tau)}\right) . \qquad (4.65)$$

Of course, instead of (4.65) the corresponding QP problem can be introduced.

Applying the NN approach two neural networks are used [29,30]. The Hopfield analog, called network N1 is applied to the analysis of Eq. (4.65) and network N2, is used for the computation of control vector $z^{(\tau)}$. The computations are repeated for each value $\tau = 1,\ldots,T$ so for the sake of simplicity the index τ is omitted in what follows.

The weights of network N2 are computed according to the following formula:

$$v_{rj}^{(k+1)} = v_{rj}^{(k)} + \eta^{(k)}\left(s_r^* - \sum_{l=1}^{n} v_{rl}^{(k)} s_l^{(k)}\right) s_j^{(k)}, \quad \eta^{(k)} > 0. \tag{4.66}$$

The control parameters z_v for $v = 1,\ldots,m$ are computed in a similar way:

$$z_v^{(k+1)} = z_v^{(k)} + \mu_v^{(k)} \sum_{q=1}^{Q}\left(s_q^* - \sum_{l=1}^{n} v_{pl}^{(k)} s_l^{(k)}\big|_q\right) s_q^{(k)}, \quad \mu_v^{(k)} > 1, \tag{4.67}$$

In formula (4.67) summing $\sum_q$ denotes that the learning term is extended over all the parameters related to the v-th value of control parameters z_v, e.g. index v can correspond to the v-th finite element of the discretized structure, $s_l^{(k)}\big|_q$ and $s_q^{(k)}$ are the state variable q-th components corresponding to the v-th element. The $\sum_q$ expression may be extended on an area related to z_v.

After the control parameters $z_v^{(k+1)}$ have been computed for the $(k+1)$ step, the matrix $M\left(z_v^{(k+1)}\right)$ and vector $q\left(z_v^{(k+1)}\right)$ can be calculated and new values $x_j^{(k+1)}$ are computed by means of Eqs (4.25). Then the state vector components $s_r(x_j)$ are computed.

The computation is continued until the convergence criterion, related to (4.64), is satisfied:

$$\max_{\tau,s}\left\| s_r^{*(\tau)} - s_r^{(\tau)} \right\| < \delta, \tag{4.68}$$

where: δ – prescribed constant.

4.5.4. Tracing of yield surfaces

The experiments by Shiratori and Ikegami [33] were taken as a base for computing of yield surfaces. The experiments were carried out on a cross-shaped specimen shown in Fig.4.20a. The specimen of thickness $h = 1\ mm$ was subjected to biaxial tension along propotional and combined loading paths, cf. Fig.4.20b.

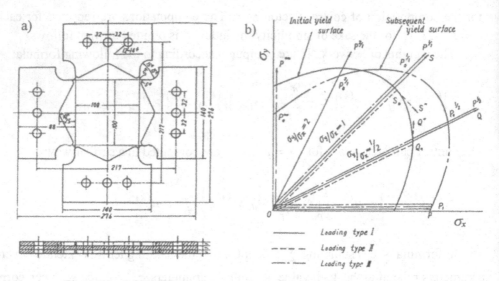

Fig.4.20: a) Biaxial tension specimen, b) Loading paths

After prestressing, unloading and reloading points $\{\sigma_x, \sigma_y\}$ were found both at the initial and subsequent yield surfaces. In Fig.4.21a such yield surfaces are shown for the prestressing path $\sigma_y / \sigma_x = 1$, cf. [33].

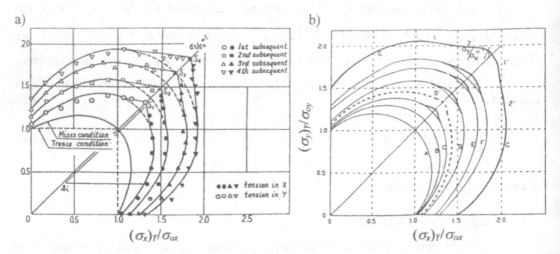

Fig.4.21: a) Experimental points at subsequent yield surfaces for prestressing $\sigma_y / \sigma_x = 1$,
b) Subsequent yield surfaces with neural prediction

For the numerical analysis it was assumed that the considered square of specimen is made of elastic material with four orthotropy coefficients which are treated as components of control vector $z = \{z_v\} = \{\alpha_{11}, \alpha_{12} = \alpha_{21}, \alpha_{33}\} = C$.

In order to get a more reliable approximation of the anisotropic elastoplastic problem an intermediate yield surface was computed between the experimental yield surfaces which are assumed to be the elliptic paraboloid surfaces. After the initial „rough" prediction the neural nets N1 and N2 are used to solve the parameter identification problem, as it has been discussed in the previous Point 4.5.3. The predictions σ_{pl} are corrected by means of the following updated formula:

$$\sigma_{pl}^{(k+1)} = \sigma_{pl}^{(k)} + \eta^{(k)} \sum \left(z(\sigma_{pl}^{(k)}) - z_o \right), \qquad\qquad (4.69)$$

where: $z(\sigma_{pl}^{(k)}) = C^{(k)}$ – elastic coefficients obtained from network N2 and $z_o = C^{(o)}$ – elastic coefficients computed on the basis of experimental results, Σ – extension of formula (4.67) over all the elements of specimen square, $\eta^{(k)}$ – a scale factor corresponding to the learning of unsupervised learning.

In Fig.4.21b the curves A, B, D and E are evaluated on the basis of experimental points shown in Fig.4.21a. On the basis of these yield surfaces the internal surface D' and the external one G are computed applying the algorithm sketched above. A slight non-convexity (e.g. parts 1-2 and 1'-2' of G) is observed at the computed yield surfaces. The larger the stresses the more visible is this lack of convexity. More detailed discussion associated with other loading paths is given in [29]

4.6. GENERAL CONCLUSIONS

The problems discussed above lead to the following conclusions:

1. Neural procedures can be efficiently used for the analysis of material nonlinear problems, especially for the analysis of elastoplastic constitutive equations. Such procedures can be successfully implemented in the hybrid programs using standard finite difference or finite element programs.

2. The Panagiotopoulos approach, which is based on the QP formulation and application of the Hopfield-Tank network, can be efficiently applied to the analysis of bilateral and unilateral problems related to classical crack models and elastoplastic structures.

3. The application of neural networks to the analysis of inverse problems in mechanics has a very promising prospect. This approach was used to the analysis of parameter identification problems for material characteristics.

The proposed approaches were illustrated with selected examples and need further theoretical and numerical investigations.

The co-author of this Chapter, Professor Panagiotopoulos died in August, 1998. His genuine ideas originated a new approache to solving difficult mechanical problems and that is why those ideas seem to be worth developing in near future.

ACKNOWLEDGEMENT

Sections 4.2 to 4.6 were also written by Z.Waszczyszyn. He was in intimate contact with Professor P.D. Panagiotopoulos before his decease. Prof. A.V. Avdelas from the Aristotle University, Thessaloniki, Greece is cordially acknowledged for his remarks to the manuscript of these Sections.

Financial support by the Polish Committee for Scientific Reserarch, Grant No 8 T11 022 14 „Artificial neural networks in structural mechanics and bone mechanics", is gratefully acknowledged.

REFERENCES

1. Hajela, P. and Berke, L.: Neurobiological computational models in structural analysis and design, Comp. & Stru., 41 (1991), 657-667.

2. Szewczyk, Z.P. and Noor, A.K.: A hybrid neurocomputing/numerical strategy for nonlinear analysis, Comp. & Stru., 58 (1996), 661-667

3. Ghaboussi, J., Garret, J.H. Jr and Wu, X.: Knowledge-based modeling of material behaviour with neural networks, J.Eng. Mech., 117 (1991), 132-153.

4. Jadid, M.N. and Fairbairn, D.F.: Neural-network applications in predicting moment-curvature parameters from experimental data, Eng. Applic. Artif. Intell., 9 (1996), 308-319.

5. Mucha, G. and Waszczyszyn, Z.: Hybrid neural-network/computational program for bending analysis of elastoplastic beams, in: Proc.XIII Polish Conf. on Comp. Meth. in Mech., Poznań 1994, Vol, 3, 949-954.

6. Yamamoto, K.: Modeling and hysteretic behaviour with neural network and its application to nonlinear dynamic response analysis, in: Applic. Artif. Intell. Eng., Proc. 7th Intern. Conf., 1992, 475-484.

7. Waszczyszyn, Z., Pabisek, E. and Mucha, G.: Hybrid neural network/ computational programs to the analysis of elastic-plastic structures, in: Discretizations Methods in Structural Mechanics II (Eds. H.Mang and F.G. Rammerstorfer), Kluwer Acad. Publ., Dortrecht 1999, 189-198.

8. Waszczyszyn, Z. and Pabisek, E.: Hybrid NN/FEM analysis of the elastoplastic plane stress problem, Com. Assisted Mech. Eng. Sci., 1999, (in press)

9. Pabisek, E. and Waszczyszyn, Z.: Hybrid, BPNN/FEM analysis of the elastoplastic plate bending, in: Proc. of the 4th Conf. on Neural Networks and Their Applications, Zakopane-Częstochowa, Poland, 1999, (in press)

10. Krzyś, W. and Życzkowski, M.: Elasticity and Plasticity (in Polish), PWN, Warsaw 1962.

11. Ramm, E. and Matzenmiller, A.: Computational aspects of elasto-plasticity in shell analysis, in: Owen, D.R.J. et al. (Eds.), Computational Plasticity – Models, Software and Applications, Pineridge Press, Swansea 1987, P.I, 711-734.

12. Simo, J.C. and Taylor, R.L.: A return mapping algorithm for plane elastoplasticity, Int. J. Num. Meth. Eng., 22 (1986), 649-670.

13. SNNS - Stuttgart Neural Network Simulator, User Manual, Version 4.1, Institute for Parallel and Distributed High Performance Systems, Rep. No 6/95, Univ. of Stuttgart, 1995.

14. Waszczyszyn, Z., Cichoń, C. and Radwańska, M.: Stability of Structures by Finite Element Methods, Elsevier, Amsterdam 1994.

15. Owen, D.R.J. et al: Stresses in a partly yielded notched bar – an assessment of three alternative programs, Int. J. Num. Meth. Eng., 6(1973), 63-73.

16. Hopfield, J.J. and Tank, D.W.: "Neural" computation of decisions in optimization problems, Biol. Cybern., 52(1985), 141-152.

17. Kortesis, S. and Panagiotopoulos, P.D.: Neural networks for computing in structural analysis – methods and prospects of applications, Int. J.Num. Meth.Eng., 36 (1993), 2305-2318.

18. Theocaris, P.S. and Panagiotopoulos, P.D.: Neural networks for computing in fracture mechanics – methods and prospects of applications, Comp. Meth.Appl. Mech. Eng., 106 (1993), 213-228.

19. Avdelas, A.V., Panagiotopoulos, P.D. and Kortesis, S.: Neural networks for computing in the elastoplastic analysis of structures, Meccanica, 30 (1995), 1-15.

20. Panagiotopoulos, P.D.: A nonlinear programming approach to the unilateral contact- and frictional–boundary value problem in the theory of elasticity, Ing.-Archiv, 44 (1975), 421-432.

21. Pu, S.L., Hussain, M.A. and Lorensen, W.E.: The collapsed cubic isoparametric element as a singular element for crack problems, Int.J.Num. Meth. Eng., 12 (1978), 1727-1742.

22. Künzi, H. and Krelle, W.: Nichtlineare Programierung, Springer, Berlin 1962.

23. Maier, G.: Incremental plastic analysis in the presence of large displacements and physical instabilizing effects, Int.J.Solids Stru., 7 (1971), 345-372.

24. Maier, G.: A matrix structural theory of piecewise linear elastoplasticity with interacting yield planes, Meccanica, 8 (1970), 54-64.

25. Panagiotopoulos, P.D., Baniotopoulos C.C. and Avdelas, A.V.: Certain propositions on the activation of yield modes in elastoplasticity and their applications to deterministic and stochastic problems, ZAMM, 64 (1984), 491-501.

26. Borkowski, A.: Analysis of Skeletal Structural Systems in the Elastic and Elastic-Plastic Range, PWN Polish Sci. Publ. and Elsevier, Warszawa/Amsterdam 1988.

27. Keller, E.L.: The general quadratic optimization problem, Mathematical Programming, 5 (1973), 311-337.

CHAPTER 5

NEURAL NETWORKS IN ADVANCED COMPUTATIONAL PROBLEMS

B.H.V. Topping, A.I. Khan, J. Sziveri, A. Bahreininejad,
J.P.B. Leite, B. Cheng and P. Iványi
Heriot-Watt University, Edinburgh, UK

ABSTRACT

It this chapter we discuss the use of parallel processing for the training of back propagation neural networks. This leads to a discussion of how neural networks and genetic algorithms may be utilised for preprocessing large finite element problems for parallel or distributed finite element analysis. This preprocessing is the partitioning of the finite element mesh into subdomains to ensure load balancing and minimum interprocessor communication during the parallel finite element analysis on a MIMD distributed memory computer.

5.1 Introduction

Design in many branches of engineering requires linear and non-linear finite element analysis. An ever increasing requirement to explore more design possibilities results in large scale finite element computations which may be part of an optimisation process or a simulation of an artifact's behaviour. Parallel and distributed computing permits the engineer to undertake the finite element analysis in a considerably shorter time, reducing the time between development of design concept and production of the final artifact.

Application of parallel and distributed computing to finite element analysis requires the development of parallel algorithms. For these algorithms to be applied effectively then considerable pre-processing of the finite element model is required to ensure that the analysis is efficiently undertaken in parallel. New and emerging technology in this area includes neural networks and genetic algorithms. This chapter briefly discussess parallel and distributed computing and then parallel neural networks and parallel genetic algorithms. Finally it is demonstrated how these techniques may be utilised in the preprocessing for a parallel non-linear transient finite element analysis.

The concepts discussed here result in the generation of finite element meshes in parallel permitting the finite element discretization to be generated in parallel without ever having to be assembled on a master or controlling processor.

Structural analysis and design for large scale problems requires considerable computational effort. The implementation of parallel algorithms for structural engineering problems have been the subject of much research recently [2]. Parallel algorithms may be generally classified as being of one of three types:

- Processor Farming: In this type of algorithm each processor in the network executes code in isolation from all other processors except for a Master or Root processor. This type of parallelisation is also called Independent Task, Task Farming or Event Parallelism. With Processor Farming interprocessor communication is not generally permitted.

- Geometric Parallelism: In which each processor executes the same codes on data corresponding to a sub-region or sub-domain being simulated or processed. Interprocessor communication is possible to permit the exchange of boundary data between neighbouring processors representing connections between the subdomains.

- Algorithmic Parallelism: In which each processor is responsible for part of the algorithm and the data passes through each processor in turn. Different parts of the algorithm are executed on each processor. The processors represent a conveyor belt on which computation is performed. The computational load on each processor must be the same or balanced if a bottleneck is not to develop.

With processor farming and geometric parallelisation the processors must all complete their tasks at the same time if processors are not to be left idle waiting for others to complete their tasks. This requires load balancing of the computations which is an important aspect of parallel computations which will be discussed later. Parallel processing provides increased computational power for structural analysis by using a number of different strategies. Generally these strategies may be classified as follows:

- The analysis problem may be sub-divided by geometrically dividing the idealization into a number of sub-domains. Since each of these domains will be subject to the same instructions hence Processor Farming or Geometric Parallelisation techniques may be used. This is described as an explicit domain decomposition approach.

- Alternatively the system of equations for the whole structure may be assembled and solved in parallel without recourse to a physical partitioning of the problem. This is described as an implicit domain decomposition approach.

The explicit domain decomposition approach requires the finite element mesh to be split into a number of sub domains using a domain decomposition algorithm. Domain Decomposition approaches may be described as divide and conquer algorithms since their object is to divide the larger problem into a series of smaller subproblems. There are numerous techniques for domain decomposition however two basic approaches are common:

1. Iterative Techniques: are used to solve the whole problem iteratively and information concerning nodes common to two or more domains must be communicated between processors after each iteration.

2. Sub-structuring techniques have been established for over thirty years. In this approach the sub-domains are treated as complex structural elements and the formulation for each sub-domain is carried out by only taking into account the degrees of freedom of the boundary nodes. Once the displacements of the boundary nodes have been determined the displacements within each substructure may be determined independently.

This chapter concentrates on the iterative approach which is appropriate for nonlinear transient finite element analysis. For the implementation of a parallel algorithm it is necessary that the given domain of the problem is discretized into a finite number of subregions or subdomains. In this regard it is important that the domain decomposition algorithm should be able to: handle irregular mesh geometries of arbitrarily shaped domains to make the method completely general; and to minimize the interface problem by providing partitioning interfaces which deliver minimum boundary node connectivities. This aids the reduction of the magnitude of the boundary problem and in physical terms reduces the communication overheads in the parallel computing system; while the subdomains should carry approximately equal amounts of computational load. This is again to satisfy the physical requirement of the system ensuring that all the processors finish their work at about the same time and that the controlling processor is not forced to wait on account of a lagging processor. Mesh decomposition algorithms are discussed in detail elsewhere [2].

The range of parallel computer architectures is discussed with respect to structural design in [1, 2] but the dominant computer architecture is currently MIMD (multiple instruction multiple data) with distributed memeory. This classification corresponds to a wide range of supercomputers, high performance multiple processor computers and loosely coupled networks of workstations.

5.1.1   Distributed and Heterogeneous Computing

The role of distributed computing is increasing because personal computers and work-stations may now be networked providing engineers with more powerful integrated systems, without the additional expense of high performance computers. These networks may be local area networks (LANs) or wide based area networks (WANs). They usually utilise the TCP/IP protocol with various types of network carrier including: thin ethernet coaxial cable; fibre optics; twisted pair cable; and ISDN links. These networks are usually heterogeneous in the sense that all the connected processors will not usually be of the same type or power. Some nodes of the network may themselves be MIMD or SIMD (single instruction multiple data) computers which represent high performance compute nodes within the network. It is important to ensure that any computational task is allocated to the appropriate processor within the heterogeneous network.

PVM

Parallel Virtual Machine (PVM) [3] is a portable message passing programming system designed to link separate Unix host machines to create a virtual machine which is a single manageable computing resource. The PVM system is composed of two parts. The first is a daemon program which runs on all machines that are part of the virtual machine. This permits the user to run a PVM application at a Unix prompt from any of the machines. The second part of the system is a library of PVM interface routines. This library contains user callable routines for passing messages, spawning processes, coordinating tasks and modifying the virtual machine. Application programmes can be written in a mixture of C, C++ and FORTRAN but must be linked with the PVM library.

Message Passing Interface

Message Passing Interface (MPI) [4] is a specification for a library of routines to be called from C and FORTRAN programs. This system provides an API (Application Programming Interface) which permits a standard to develop parallel programmes with standard message passing. This interface is a standard and a carrier such as PVM or

P4 is required for distinct implementations if the operating system does not support the message passing. Both native and layered implementations are known for a variety of systems.

5.2   Parallel Neural Networks

Artificial Neural Networks (ANNs) are computational models which attempt to mimic the learning function of the brain. The brain is the most complex system comprising billions of neurons. Each neuron is a processing unit which receives, processes and sends information. A biological neuron is made up of three parts: the *cell body*; the *axon*; and the *dendrites*. A typical biological neuron is shown in Figure 5.1. Signals travel through the axon of other neurons thus reaching the dendrites of another neuron. The signals then travel through a junction between the axon and the dendrites called the *synapse*. The response of a neuron depends on several biological and chemical factors corresponding to the synapses and the receiving neuron. A neuron may fire a signal if the magnitude of the signal is strong enough to activate it. The synaptic efficiency or strength is modified to adjust to the received signal. The brain is said [5] to learn when the synapses adjust themselves to accommodate to receive new signals.

Artificial neural networks are composed of a set of neurons or processing units which are connected together by means of connecting weights. ANNs which are structured to learn and generalize so that the network may learn by continuous adjustment of the weights of the connections.

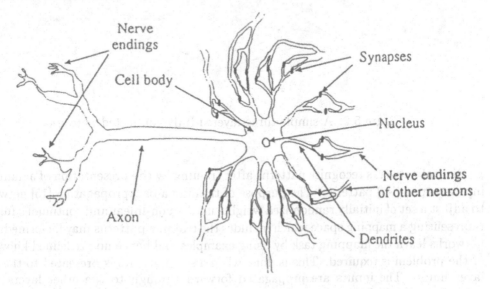

Figure 5.1: Idealized representation of a biological neuron.

Application of ANNs have received particular attention in the field of robotics,

vision recognition, military and space exploration [6, 7, 8, 9, 10, 11]. One of the drawbacks of ANNs is that they are computationally expensive. This drawback may largely
depend on the size of a network or the quantities of data involved. Despite the efforts
that have been taken to improve the training algorithms to make them more efficient
particularly in terms of computation time [12, 13, 14], the greatest scope for improving
the computational efficiency may lie with the parallel nature of ANNs algorithms. This
nature enables the algorithms to be implemented on parallel architectures.

5.2.1   Backpropagation Neural Networks

A backpropagation (BP) network has a multilayered topology which is shown in Figure 5.2. TH BP network consists of a set of neurons or units organized into a sequence
of layers with full or patterned connecting weights between the layers. Biases may
also be employed which act as weights from imaginary neurons that have an output
of one at all times. Typically only two layers terminate a network: the input layer to
which data is presented for the network, and the output layer where the response of
the network to a given input is received. Layers other than these two are called hidden
layers.

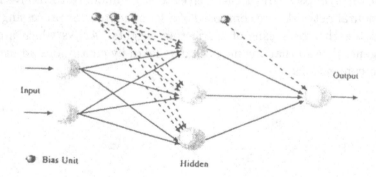

Figure 5.2: A simple three-layered fully connected network.

Neural networks recognise patterns after training by the presentation of a number of
input-output (i-o) patterns. The purpose of training a backpropagation [15] network is
to adjust a set of initially randomized weights until a non-linear and continuous function
representing a mapping space which includes the training patterns may be formed. Such
networks learn the mapping task by using examples and hence no predefined knowledge
of the problem is required. This is done when a set of inputs is presented to the input
layer units. The inputs are propagated forward through to the other layers where
they are multiplied by the associated weights. The sum of these products within each
receiving unit is subjected to a transfer function which is non-linear, non-decreasing
and differentiable. Commonly a sigmoid function is adopted as the transfer function.

The result is an output which becomes the input to the next layer. Figure 5.3 shows the input and output from a unit.

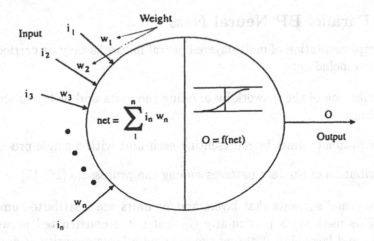

Figure 5.3: The operation of a single artificial neuron.

At the output layer, the computed output is compared with the desired output and an error value is computed. This error value together with the derivative of the transfer function are used to compute the error signal(s) for the output layer unit(s). The error signal(s) from the output layer unit(s) propagate back in order to calculate the error signal(s) for the previous hidden layer(s) [15]. Once all the appropriate error signals have been computed they are used to calculate the delta-weights representing the quantity for which the corresponding weights should change. The learning process corresponds to performing gradient descent on a surface in weight space whose height at any point in weight space is equal to the measured error. The training process is repeated for all the training patterns until a desired Root Mean Square (RMS) error level over all training patterns is achieved [16]. The RMS value of the training is given by:

$$RMS_{err} = \sqrt{\frac{1}{PN_{output}} \sum_{p}^{P} \sum_{output}^{N_{output}} (t_p - o_p)^2} \tag{5.1}$$

where P is the total number of training patterns, p is a single training pattern, N_{output} is the number of units in the output layer, t_p is the desired output in the training data and o_p is the output of the network.

Training may be undertaken using either single or batch pattern training where:

- Single Pattern Training: The weights are modified after each i-o pattern has been presented to the net.

- Batch Pattern Training: The weights are changed after the presentation of a batch or all i-o patterns [16, 17].

5.2.2   Parallel BP Neural Nets

Parallel implementation of multi-layered neural networks may be carried out in a number of forms including:

- distribution of the network by dividing the units and or layers amongst the processors;

- distribution of units by representing each unit with a single processor; and

- distribution of i-o data patterns among the processors [16, 17].

The first method suggests that layers and/or units are distributed amongst the processors. This methods is particularly efficient if the constructed network consists of many units and layers but if during the learning a large quantity of data has to travel through the pipeline then the method will be inefficient.

The second paralellisation method is to represent each unit with an individual processor. Except for small network structures an efficient implementation of this method may not be possible. This is due to the limited number of hardware links available in some parallel computers and the difficulty of undertaking point to point routing. Any point to point routing implementation will not contribute to the efficiency because the communication overhead would significantly increase.

If the network structure is not large but a large number of training patterns is to be used then the third method is likely to be more efficient. An example of parallel training of BP nets using the third approach may be found in references [16, 17] These references examine the parallel training of BP nets using single and batch training approaches. It was concluded [16] that the single pattern implementation requires a much higher degree of inter-processor communication. Batch pattern training may appear more appropriate for parallel training, however a much slower rate of convergence generally results with this form of training. The overall efficiency may be impaired, although the training patterns may be presented to the network a far greater number of times than would required with single pattern training.

5.2.3   Parallel Mean Field Annealing

Combinatorial optimization problems arise in many areas of science and engineering. Most computational solution methods that have been developed which generally yield good solutions to these problems rely on heuristics of some form. ANNs make use of highly interconnected networks of simple neurons or units which may be programmed to find approximate solutions to these problems [18, 19, 20]. They are also highly parallel systems and have significant potential for parallel hardware implementations.

The origin of the optimization neural network goes back to the work by Hopfield and Tank [21] which was a formulation to solve the Traveling Salesman Problem (TSP). The Hopfield network is a feedback type of neural network where the output(s) from a processing unit is (are) fed back as the input(s) of other units through their interconnections. Figure 5.4 illustrates a typical feedback architecture and Figure 5.5 shows a simple representation of a Hopfield network.

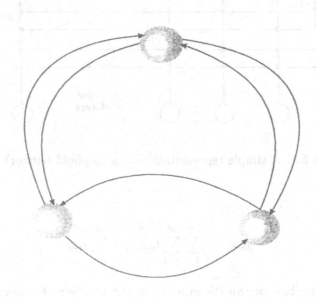

Figure 5.4: A typical feedback/recurrent architecture.

Following the poor performance of Hopfield networks in determining valid solutions for the TSP problem there followed considerable research effort to improve the performance of this type of network and to find ways of applying it to other optimization problems. At about the same time of the emergence of Hopfield networks, a new optimization method called *Simulated Annealing* [22] was developed. This technique provides a method for finding good solutions to most combinatorial optimization problems, however the algorithm takes a long time to converge. To overcome this problem, Mean Field Annealing (MFA) networks were proposed as an *approximation* to simulated annealing. MFA is a deterministic approach which essentially replaces the discrete degrees of freedom in a simulated annealing problem with their average values as computed by the Mean Field Approximation. The components of MFA networks include:

- **Hopfield Network:** NP problems may be mapped onto the Hopfield discrete state neural network using the Liapunov function which is defined as:

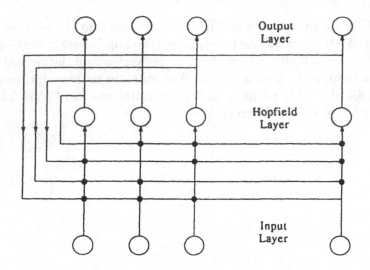

Figure 5.5: A simple representation of a Hopfield network.

$$E = -\frac{1}{2}\sum_{i=1}^{N}\sum_{j=1}^{N}t_{ij}s_{i}s_{j} + \sum_{i=1}^{N}I_{i}s_{i} \qquad (5.2)$$

where: I_i is the bias of the ith unit; N is the number of processing units; s_i and s_j represent the state of the units i and j in the network connected through the weight t_{ij}. It is assumed that the t_{ij} matrix is symmetric and has no self-interaction (i.e. $t_{ii} = 0$).

To minimize E on the energy landscape, the network state is modified asynchronously from an initial state by updating each processing unit using the updating rule:

$$s_i = sgn(\sum_{j=1}^{N}t_{ij}s_j - I_i) \qquad (5.3)$$

where the output of s_i of the ith unit is fed to the input s_j of the jth unit asynchronously by the connection t_{ij} where there are N connections from other j processing units to the ith unit. The symmetry of the matrix t_{ij} with zero diagonal elements enables E to decrease monotonically with the updating rule.

In optimization problems, the concept is to associate the Liapunov function given in equation 5.2 with the problem objective function by setting the connection weights and input biases appropriately.

- **Simulated Annealing:** simulated annealing is a probabilistic hill-climbing algorithm which attempts to search for the global minimum of the energy function. It carries out uphill moves in order to increase the probability of producing solutions with lower energy. The method carries out random search in order to find new configurations using the Boltzmann-Gibbs distribution:

$$P(\vec{S}) = \frac{e^{-\frac{E(\vec{S})}{T}}}{Z} \qquad (5.4)$$

where T is the temperature of the system and Z is called the partition function which is of the form:

$$Z = \sum_{(\vec{S})} e^{-\frac{E(\vec{S})}{T}} \qquad (5.5)$$

However simulated annealing involves stochastic search for generating new configurations. In order to reach good solutions, a large number of configurations may have to be searched which involves slowly lowering the temperature and is therefore a very CPU time consuming process.

- **Statistical Spin Glass Models for Neural Networks:** the simplest model of a physical system is the two-state Ising Spin-Glass model which allows the spin variables to take values of +1 or −1, or 0 or 1. A Potts model [23, 24, 25, 26] is the generalization of the two-state Ising model which has Q equivalent ground states where all the spins are identical but may take any one of the Q values.

The application of the two-state Ising model to partitioning of: graphs [27, 28]; and adaptive finite element meshes is demonstrated in [19]. However, the present chapter demonstrates the parallel implementation of the multi-state glass model which is used for the mesh partitioning problem. Thus for a system of N neurons where each neuron can occupy Q discrete states and the state space of the neurons is restricted such that exactly one neuron at every Q state is allowed to be on. Using the binary values of 1 and 0 representing whether the state of a neuron is on or off, the neuron restriction terminology may be written as:

$$\sum_{a}^{Q} S_{ia} = 1 \qquad (5.6)$$

thus a neuron i can be on (1) in only one of the Q possible states and off (0) in the remaining states.

Peterson and Söderberg [29] presented a new method for mapping optimization problems onto neural networks. By considering the equations 5.4 and 5.5, it has been shown [29, 30] that the discrete sum of the partition function may be replaced by an integral over the continuous variables u_i and v_i and to evaluate (approximately) the integrand of this integral at the saddle point, therefore:

$$Z = C \prod_{j=1}^{N} \int_{-\infty}^{\infty} dv_j \int_{-i\infty}^{i\infty} du_j e^{-E'(\vec{V},\vec{U},T)} \tag{5.7}$$

where C is a constant and $\prod \int \int$ refers to multiple integrals. For the two-state Ising model E' may be given in the form:

$$E'(\vec{V},\vec{U},T) = E(\vec{V})/T + \sum_{i=1}^{N} u_i v_i - \log(\cosh u_i) \tag{5.8}$$

and for the Q-state Potts model E' may be given in the form:

$$E'(\vec{V},\vec{U},T) = E(\vec{V})/T + \sum_{i=1}^{N} u_i v_i - \sum_{i=1}^{N} \log \sum_{k=1}^{Q} e^{u_{ik}} \tag{5.9}$$

and the multiple integrals may be determined using a saddle point expansion of E' which involves the mean field approximation that is found in references [28, 29]. The saddle point positions for the Potts model are given by:

$$u_{ia} = \frac{-\partial E'}{\partial v_{ia}} \tag{5.10}$$

$$v_{ia} = \frac{e^{u_{ia}}}{\sum_{b}^{Q} e^{u_{ib}}} \tag{5.11}$$

where $\sum_{a}^{Q} v_{ia} = 1$.

Furthermore, the continuous variables, v_i, are used as approximations to the discrete variables at a given temperature (i.e. $v_i \approx \langle s_{ia} \rangle_T$), thus the final value of v_{ia} approximates whether the value of s_i is 1 or 0.

Equations 5.10 and 5.11 are the iterated asynchronously. This is based on updating the value of only one v_{ia} after each time-step $(t + \Delta t)$.

Topping and Bahreininejad [20] demonstrated the use of neural networks in efficiently partitioning adaptive unstructured finite element meshes. The problem of partitioning meshes may be mapped onto a MFA Potts neural network and may be defined by an objective function in the form of the Hopfield energy function given in equation 5.2. The Hamiltonian for the partitioning is therefore given by:

$$E(\vec{S}) = -\frac{1}{2} \sum_{i}^{N} \sum_{j}^{N} \sum_{a}^{Q} t_{ij} \vec{s}_{ia} \vec{s}_{ja} \tag{5.12}$$

where N is the number of elements and Q is the number of states which represents the number of partitions. This is carried out by assigning a binary unit of $s_{ia} = 1$ or $s_{ia} = 0$ to each element of the mesh defining which partition the element is to be assigned. The connectivity of the triangular elements is encoded in the t_{ij} matrix in the form of:

$$
t_{ij} = \begin{cases} 1 & \text{if a pair of elements are connected by an edge} \\ 0 & \text{otherwise} \end{cases}
$$

With this terminology the minimization of equation 5.12 will maximize the connectivity within partitions while minimizing the connectivity between the partitions. This has the effect of maximizing the boundaries and therefore using this term alone as the cost function forces all the elements to move into one partition, thus a penalty term is applied which measures the violation of the equal sized partition constraint [28, 29]. Therefore the neural network Liapunov function for the mesh partitioning problem is of the form of:

$$
E(\vec{S}) = -\frac{1}{2}\sum_i^N \sum_j^N \sum_a^Q t_{ij}\vec{s}_i\vec{s}_j + \frac{\alpha}{2}\left(\sum_i^N \sum_j^N \sum_a^Q \vec{s}_{ia}\vec{s}_{ja}\right) \tag{5.13}
$$

where α is the *imbalance* parameter which controls the balance between the partitions. The second term in equation 5.13 ensures load balancing of the partitions and will be zero if the partitions are correctly balanced. From equations 5.10 and 5.13 therefore:

$$
u_i = \frac{\sum_i^N \sum_j^N \sum_a^Q (t_{ij} - \alpha)v_{ja}}{T} \tag{5.14}
$$

The saddlepoint equations 5.11 and 5.14 are used to compute a new value of v_{ia} asynchronously. Initial values for the **V** matrix are assigned using $\frac{1}{Q}$ plus a small noise factor increment. The temperature is lowered using a cooling factor (typically 0.95) after one complete iteration of equations 5.11 and 5.14.

The Q spins associated with a particular element i can be looked upon as the probabilities of the element being assigned to each of the partitions and therefore from $\sum_a^Q v_{ia} = 1$, an element can occupy only one of the partitions.

The MFA Potts network was formulated for a parallel environment [20]. The parallel implementation of the annealing process was carried out using a pipeline network of transputers. The annealing process is applied using the asynchronous updating method. Figure 5.6 shows the flowchart of the parallel algorithm.

First the values of the initial parameters and the element information are sent from the Master task to the Worker tasks. The number of Workers is equal to the number of required partitions. Next the weight matrix t_{ij} which represents the elements connectivity is setup on all Worker processors. The initial spin matrix, **V**, is generated on the Master and is distributed vectorially between the Workers. This means, for example, if the initial spin matrix is made of three columns representing three partitions, then each column is sent to a Worker.

Once the annealing starts, each Worker computes the value of equation 5.14. Each Worker then sends its result to the Master where equation 5.11 is computed and the resulting v_{ia} values are sent back to the appropriate Workers. This procedure is carried

Master   Worker

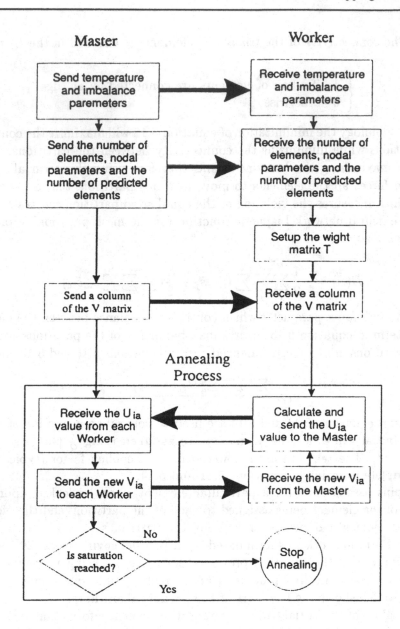

Figure 5.6: Flowchart showing the parallel Potts annealing.

out until all the elements have been accounted for the annealing. The temperature is reduced on the Master task. Once a desired level of saturation has reached (typically 0.999), the annealing stops. The saturation level is given by:

$$sat = \frac{(\sum v_{ia}^2)}{N} \tag{5.15}$$

Thus at the end of annealing, the Master has all the optimized spin values in the form of a V matrix. This matrix represents the final partitioning solution.

The sequential MFA Potts neural network carried out the partitioning on a single transputer. The parallel code was run on a number of processors corresponding to the number of partitions required. To assess the efficiency of the parallel code, both the parallel and sequential codes were run for a total of 100 iterations on the problem domain shown in Figure 5.7. The comparison for the 100 iterations are given in Table 5.1. References [19, 20] provide more details on partitioning finite element meshes using MFA neural networks.

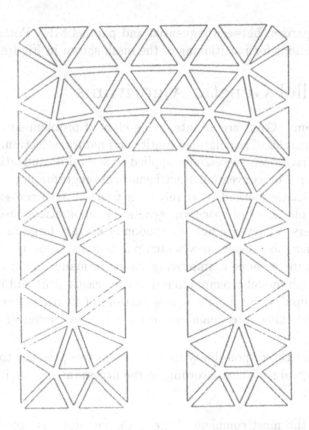

Figure 5.7: The domain tested for the sequential-parallel MFA partitioning.

The initial coarse mesh shown in Figure 5.7 was partitioned into seven partitions using the MFA Potts method. Figure 5.8 shows the partitioned mesh based on the coarse mesh and Figure 5.9 shows the partitioned mesh after remeshing.

No. of Partitions	Computation Time (sec)	
	Sequential Code	Parallel Code
1	6	6
2	13	8
3	20	9
4	27	11
5	33	12
6	40	13
7	47	15
8	54	17

Table 5.1: Comparison between sequential and parallel MFA Potts method run for a total of 100 iterations for partitioning of the mesh shown in Figure 5.7.

5.3 Parallel Genetic Algorithms

Genetic Algorithms (GAs) are an interesting class of problem solving technique that combines the principles of population genetics and natural selection. These approximation algorithms have been successfully applied to several optimization problems which are difficult to solve by conventional mathematical programming.

GAs are stochastic methods that rely on a trade-off between exploitation of good solutions and exploration of space via genetic recombination based on probabilistic control parameters. They are able to randomly sample large areas of the problem search space. Then, they evolve new search points based upon the performance of the old search points in the hope of improving the performance of the overall search.

The GA's search model attempts to mimic the genetic drift and Darwinian strife for survival. By manipulating symbolic representations of solutions, evolution in canonical GAs, is performed using three main operators: selection, crossover and mutation.

- Selection has the responsibility to impose selective pressure towards convergence exploiting good solutions according to the fitness rates of their objective function evaluations.

- Crossover, the most complete of the GA's operators, is able to evolve solutions through combination and inheritance of good features from different solutions. Crossover is basically responsible for local optimization using elements of direct search.

- Mutation promotes diversity at random and explores more global areas of the solution space providing a mechanism to escape from local optima.

Figure 5.8: The initial coarse mesh divided into seven partitions using the predictive MFA Potts neural network. The initial mesh has 75 elements.

Compared to conventional optimization techniques, GAs are population based instead of point-based, i.e. they attempt to evolve complex systems concurrently rather than to develop one and refine it. The implicit parallel properties gained by evolving a population of points in the search space concurrently [33, 34] suggests that GAs have a natural mapping onto parallel architectures.

Although canonical GAs offer great potential for parallel implementations they also employ global information and a centralised control mechanism which constrains the parallelisation to a certain level. Driven usually by the hardware availability, a variety of schemes for parallelising GAs have recently been proposed. In an attempt to increase the degree of parallelisation, reduce communication overheads and maximise efficiency of hardware resources, the majority of these parallel implementations introduce differences in structure [35, 36, 37] when compared to the canonical GA, especially in the areas of population, mating and generation models.

When significant modifications to the algorithm in the parallel model are required

Figure 5.9: The final remeshed subdomains divided into seven partitions using the predictive MFA Potts neural network. The final mesh has 1795 elements.

in order to take full advantage of the hardware available, it also seems reasonable to expect different behaviour on comparison with the original sequential algorithm. A variety of parallel algorithms have been previously proposed [38, 39, 40], whose models result in enhanced properties of population dynamics. Parallel genetic algorithms may be loosely classified according to three distinguished models: global, island and cellular.

5.3.1 Global Models

Serial GAs consider the total population as a single breeding (global) unit where mates are selected at random within the population (*panmixia*) and species evolve without any external influence. Hence, all individuals have to coexist in the same habitat and be equally available to each other for mating. Thereafter, reproduction is made in pairs of mates. This mechanism suggests that GAs inherit a certain degree of parallelism, since reproduction is an independent process and can be performed in a distributed manner. However, the overall control should reside in a single processor, since information about fitness of the entire population must be available for the selection process. Hence part of the algorithm has still to be processed in a sequential fashion.

If the GA is implemented in a traditional way, at each generation, the execution takes place on a single processor until selection is performed. Various schemes may be adopted to, thereafter, process crossovers, mutations and evaluations in a distributed manner.

However, the cost of generating and selecting a large population on a single processor is generally very high compared to the cost for reproduction of a pair of individuals. Therefore, parallelisation in such a fashion is unlikely to be highly efficient unless the computational cost of the evaluations is very high.

Earlier implementations of these parallel genetic algorithms in shared memory machines (SMM) were examined by Grefenstette [41] in the early 1980s.

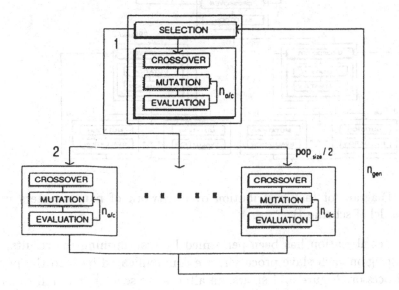

Figure 5.10: Diagram of task distribution over network of n transputers using the population model of scheme A

In the scheme A, shown in Figure 5.10, each processor including the root, is allocated two individuals for recombination. Thus the maximum number of processors, n_{max}, that may be utilised for a given population size is:

$$n_{max} = \frac{1}{2} pop_{size} \qquad (5.16)$$

If fewer than (n_{max}) processors are available then more than one pair of individuals may be task farmed to each processor for recombination. For a perfect load balancing the population size should be equal to or a multiple of the number of processors used.

In the serial algorithm used as model for parallelisation in this study, the GEBENOPT (*GEnetic Based ENgineering OPtimization Tool*) [42], there is a possibility depending on the details *a priori* specified that more than two offspring are generated per pair of parents. In Figure 5.10 this is shown by the loop on each processor generating $n_{o/c}$ offspring, where $n_{o/c}$ is the number of offspring per pair of parents.

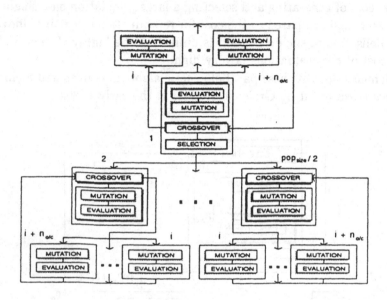

Figure 5.11: Diagram of task distribution over network of n transputers using the population model of scheme B

Once the recombination has been performed by task farming the results, i.e. the best two offspring on each slave processor, are communicated back to the population on the root processor. Figure 5.11 shows, an alternative scheme, referred to as scheme B, where again pairs of individuals are task farmed to the processors numbered 2 to $pop_{size}/2$. The crossover operation is performed on pairs of individuals by processors numbered 1 to $pop_{size}/2$ with each couple generating $n_{o/c}$ offspring which are distributed for mutation and fitness evaluation to a further $(n_{o/c} - 1)$ processors. One individual is mutated and evaluated on each of the processors and the fitness tournament takes place at the next highest level. Hence, the maximum number of processors that may be utilised for a given population size is:

$$n_{max} = \frac{1}{2}pop_{size}.n_{o/c} \qquad (5.17)$$

Since $n_{o/c}$ is always going to be two or greater then the maximum number of processors that may be utilised will be significantly greater for scheme B than for scheme A. A similar scheme was previously described in reference [43], however in that case only the fitness evaluations were performed in a distributed manner. For the scheme B similar considerations apply concerning load balancing as discussed for the scheme A.

Schemes A and B achieve higher efficiency when the fitness evaluation is computationally expensive compared to the generation of an individual. In this case, the communication overheads are also of very little significance, especially for small to medium sized populations. Previous implementations [43, 44] of these schemes have shown nearly linear speedup. Both above schemes present no structural changes from the serial model, the only difference being the hardware accelerator. Therefore, no differences in behaviour have been observed with regard to the quality of solution and the same values of control parameters for the serial algorithm may also be adopted for the parallel one.

5.3.2 Island Models

Although in theory this is not an impossibility, in nature's realm *panmixia* is only likely to occur in small and geographically isolated populations. In large populations, even though coexisting in the same continuous space, individuals tend to move and mate only within a surrounding region. This is a phenomenon known as isolation-by-distance [45]. Whether for segregated mating in neighbourhoods or for evolving semi or completely isolated subpopulations, the population is naturally subdivided.

In such a context, a parallel machine may be viewed as a global and complex environment composed of a number of independent regions. Each processor, then, corresponds to one isolated region or *island* and performs a complete GA on its own local population. These local or subpopulations may evolve in an isolated or semi-isolated regime but in any case the control and knowledge is local, i.e. the solutions or individuals are accessed and selected according to the statistics of the local population only. Hence, algorithms based in island models are coarse-grained parallel genetic algorithms (*cgp*GAs) whose parallelism comes from processing subsets of a serial population concurrently on a number of processors.

Another favourable aspect of this model is that by evolving small subpopulations the cost of some specialised operations may be substantially reduced. This is especially important for some advanced operators, present in some GAs, which may increase exponentially with the growth of the population. An example is when populations are ranked according to their fitness values. In this case, it is less expensive to rank and introduce individuals in small independent subpopulations than the entire population in a single rank.

This class of *p*GAs is suitable to solve large problems on a small network of processors. Increases in the number of processors enables the size of the subpopulations to be reduced down to a certain size, smaller than which the reproduction would be ineffective, resulting in incestuous crossbreeding and premature convergence. Thereafter, the use of a larger number of processors only results in a more representative sample which may lead to improvements in the quality of solution. This also may be viewed as an attractive feature since, in parallel GAs, the global population size may increase but this will not necessarily introduce additional computational cost.

Thus, the efficiency of these algorithms, may be directly related to the optimal serial population size. In problems where the optimal serial population size is already small, there is not much room for reductions in the population size and parallelisation using such a model is unlikely to result in substantial speedup.

The simplest scheme of this model uses identical serial GAs to evolve complete isolated subpopulations in each processor. Isolated in the sense that there is no movement of individuals among local populations. The only communication between processes is to distribute the control parameters to all nodes and gather the final results to report to the user. Shonkwiler [46, 47] presents some applications of selected problems using this *IIP* (independent and identical processing) model with isolated populations, which, he claims, achieves "superlinear speedup" to convergence. However, this claim is supported by the philosophy that the GA, as a stochastic method, bases its achievements on a number of runs whose times have to be summed for the total time.

More refined schemes introduce a certain mobility of individuals towards adjacent or nearby subpopulations. The intermixtures among subpopulations attempt to increase diversity, thus reducing the problems due to lack of schemata for recombination and hence the avoidance of premature convergence.

Topping and Leite [48] discussed two new different diffusion schemes for structural engineering problems. In the first scheme, two individuals from different subpopulations travel at the same time for pairing. Hence, the reproductive population of a processor is composed only of mates from neighbour populations, i.e. without direct contribution of the previous population. The diagram in Figure 5.12 shows the distribution of tasks over three generic subsequent nodes of a ring topology of n processors using the scheme C. In the scheme D all recombinations are made with one mate from the resident population and another mate from the neighbouring population as shown in Figure 5.13.

Once these subpopulations models address the problem of premature convergence, a remedial solution could be the hybridisation as proposed by Mansour *et* Fox [49] and by Mühlembein *et al.* [35]. The use of hill-climbing techniques to refresh converged populations could minimise the likelihood of premature convergence.

5.3.3   Cellular Models

The *cellular* models, are fine-grained parallel GAs (*fgp*GAs) composed of one large population subdivided in overlapping *neighbourhoods*, where a single individual is placed in each processor and each individual has its own particular neighbourhood. The parallel machine, in this case, may be understood as a body composed by a population of reproductive cells as shown in Figure 5.14. If the body is structured in such a way that the cells have very limited mobility, reproduction only can occur between cells placed next to each other (neighbours). Hence, in these models each individual requests a

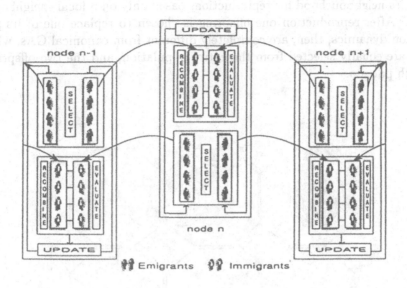

Figure 5.12: Diagram of task distribution over three generic subsequent nodes of a network of n processors in a ring topology using the subpopulation model of scheme C

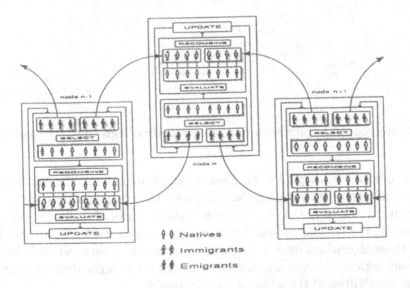

Figure 5.13: Diagram of task distribution over three generic subsequent nodes of a network of n processors in a ring topology using the subpopulation model of scheme D

mate from its neighbourhood for reproduction, based only on a local (neighbourhood) knowledge. After reproduction one offspring is chosen to replace one of its parents. The selection dynamics, then, are completely different from canonical GAs, where the two mates are equally selected from the entire population, and the two offspring may replace both parents.

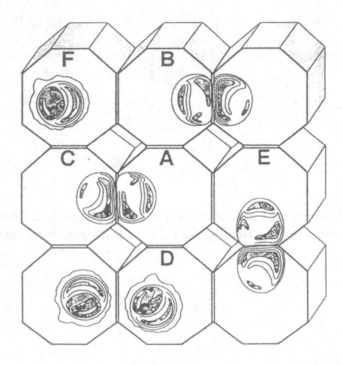

Figure 5.14: Reproductive cells with restricted mobility, segregated by their environment

As it also happens with *cgp*GAs, this class of algorithm finds its efficient applications among large problems (e.g. large populations), once the communication overheads are extremely reduced due to the small volume and short routes of messages. In fact, the high efficiency of the cellular models for some applications made researchers [50, 51] affirm that these algorithms deliver "superlinear speedup". However, this impressive speedup is not achieved by the parallelisation of the serial algorithm, but rather due to the better capabilities of the adapted parallel model.

An advantage of the *fgp*GAs is the capability to solve problems which combine very large strings (e.g. many variables and higher discretisation) massive populations within reasonable computational time and without address memory problems. An inconvenience of these models is their lack of portability since they rely on complex topologies of processors and very expensive hardware.

5.3.4 Effect of Communication on Parallel GAs

The optimal sizes of parallel populations, subpopulations and neighbourhoods are functions of the optimal serial population size and of the number of processors. Yet, the concept of optimal serial population size may be postulated as being the number of individuals required to produce sufficient schemata to solve the problem with minimum effort.

A direct distribution of a serial population over a number of processors allocates in general very few individuals to each local population. Then, if the subpopulations evolve without intercommunication, the lack of diversity induces the local populations to converge quickly but to suboptimal solutions. Therefore, parallel models using totally isolated GAs can only afford little reduction in the population size. In order to allow the use of smaller populations, communication is required to promote intermixtures among segregated populations which will restore the diversity from time to time. These intermixtures among subpopulations may occur via *migration* of individuals or via *diffusion* of features.

The main claim behind models with segregated subpopulations is the occurrence of local and global optimization concurrently. Hence by confining the search to different regions of the sample, the overlaps in the solution space may be reduced. In addition, since the competition is locally applied, different solutions may be evolved concurrently. A specific application for models using such population structure and communication mechanisms is in the optimization of multi-modal functions, in which the many local minima are of similar order of attraction to the global minimum. Competition, in canonical GAs, tends to eliminate many of the minima during the optimization process, which may eventually get stuck in a local minimum.

Migration

Migration was naturally introduced in the island models and the general idea was to send copies of individuals with high fittnesses to other processors, in order to be incorporated in their populations and make them available for selection. However, the introduction of new material may only be effective before a local population converges totally. Thereafter, a recessive genetic character will be immediately eliminated. Previous studies [52, 53, 40] have been carried out on a variety of schemes which adopt migration models. Although these migration schemes are based in this same premise, they are not necessarily equal. They may differ in the selection mechanisms, in the volume of communication (from a single individual to half subpopulation) or in the broadcast connectivity (from two processors to the entire network). The number of communication connectivities is generally constrained by the topological capabilities of the hardware. The connectivity may also be *static*, maintained constant during the entire process or it may be *dynamic*, i.e. heuristically or randomly modified throughout generations.

A potential drawback of this type of communication is that in small and semi-isolated populations, external interference, i.e. communications among populations by migration, may affect the population evolution substantially. Relatively fit individuals injected in small populations may overtake the entire population within a few generations.

Diffusion

Diffusion is the interacting mechanism of the cellular models. This is a much subtler process of intermixture among semi-isolated subsets of the population where individuals travel to visit other individuals in a small bounded area (*neighbourhood*) for mating. In Figure 5.14, for example, since cell A only knows cells B, C, D or E, it shall mate with one of these four. Therefore, since there is no contact between cells A and F, features cannot be directly transferred from one to another. However, features may be transferred from cell A to cells B or C in a generation and from one of these two last to F in a subsequent generation.

A potential drawback is the formation of colonies of layouts sharing the same features and when combined within a colony it is unlikely to produce different individuals, since there is no difference in the genetic material.

Since interprocessor comunication may be limited to four or six processors, it is not possible to physically connect all processors in the network. Thus, data sent from one processor to another may often pass through one or more intermediate processors which act as relay stations. The population structure of overlapping neighbourhoods limits the interprocessor communication to produce efficient implementations.

Later *cgp*GA implementations also use a neighbourhood structure to allow communications in two levels: migration for intermixtures within a neighbourhood and diffusion for intermixtures among neighbourhoods. This may be implemented using small neighbourhoods constituted of few local populations, thus, subsets of the local population moves to an adjacent subpopulation (neighbour) for mating.

5.3.5   Parallel Genetic Algorithms in Structural Optimization

Optimization methods based on GAs have recently been applied to various structural problems, and have demonstrated the potential to overcome many of the problems associated with gradient-based methods and mathematical programming techniques. The high computational cost of GAs, however, often limits their application to problems in which the design space can be made sufficiently small, even though GAs are most effective when the design space is large.

Although a GA-based structural optimization may require a large number of structural analysis, the optimization process is parallelisable to a high degree. For large structural optimization problems, distributed processing will enhance the efficiency of GAs by cutting the high computational cost which is the only drawback of the GAs.

In 1994, Punch *et al.* [54] showed the impact of parallelisation in the design of laminated composite structures by genetic algorithms. These authors reported a superlinear speedup using an island model GA distributed over a cluster of five SUN Sparc10 workstations connected via a local area network (LAN). In addition, the results obtained by the island model GA proved to be more refined than previous results obtained using the serial model. In the same year, Adeli and Cheng [55] used a global model pGA for the optimization of large structures such as high-rise building structures and space stations with several hundred members. In this adopted model only the evaluations were performed in a distributed manner. A maximum speedup of 7.7 times was achieved for a 35-story tower (with 1,262 elements and 936 degrees of freedom), using a shared memory machine (SMM) Cray Y-MP 8/864 with eight processors. In 1995, Adeli and Kumar [56] obtained again nearly linear speedup on the optimization of truss structures using the same global model, however in this case, distributed over a heterogeneous network of eleven IBM RS/6000 workstations. Later in the year, the same authors [57] compared the results obtained by this global model running in a distributed memory machine (DMM) with 512 processors, the CM-5, with the results previously obtained using the Cray-MP 8/864. The SMM showed a better performance than the DMM for this global model, using a same population size and number of processors. This difference in performance is expected since in SMMs the information exchange between processors is performed by means of global memory and therefore, it does not introduce communication and synchronisation overheads due to interprocessor communication. On the other hand, SMMs are, in general, provided with a small number of processors and their global memory are not able to hold data for very large problems. The use of a DMM with a larger number of processors allowed Adeli and Kumar [57] to solve very large problems such as a 147-story space tower structure consisting of 4,016 truss members and 817 nodes in an affordable computational time. Although these increase in hardware power reduces the of efficiency of the system, it provides substantial acceleration to the execution process. In addition, DMMs enable the use of larger populations which may result in more representative sample and, consequently, in more refined results. Finally, the main difference in performance between high performance computing environments such as cluster of workstations and dedicated DMMs lies also on the communication process: dedicated DMMs use faster links for interprocessor communications. Yet the heterogeneity of machines and the unpredictable multi-user occupation of networked workstation environment may introduce relative complexity in the load-balancing mechanism. Nevertheless, the cluster of workstations can in general provide a better price/performance for structural optimization problems. Topping & Leite [48] examined the applicability and relative merits of the many parallel models and different parallel environments for engineering optimization. These authors employed a design problem of a cable-stayed bridge to illustrate the differences in performance on both speed and quality of solution of different parallel GA models. They discussed the different models and the computer architectures, topologies and applications where these models are expected to produce higher efficiency.

The results from these studies shown that, specially for large engineering problems, the parallel GAs perform better than the serial algorithm both in execution speed and quality of the solution.

5.3.6 Mesh Partitioning using Genetic Algorithms and Neural Networks

In the first section of this chapter, parallel and distibuted computing were discussed. In the second and third sections parallel neural networks and genetic algorithms were discussed respectively. Finally, in this section, a parallel version of a mesh partitioning technique for unstructured adaptive planar finite element meshes is described which utilises a genetic algorithm and a neural network predictor. This method or partitioning is called the Subdomain Generation Method (SGM).

Conventional mesh-partitioning algorithms operate on the overall final mesh to obtain optimal partitions. With the SGM the overall mesh is not formed and hence a different strategy is adopted. The strategy used for the SGM relates to the adaptive mesh generation method described in references [2, 58, 59], where the adaptive refined mesh, which is the final mesh to be analysed, is generated using the following information:

- an initial coarse (background) mesh; and

- the element mesh parameters δ_e for controlling the local mesh density.

In order to practically implement the SGM the initial mesh must be partitioned into a suitable number of sub-domains such that each sub-domain of the refined mesh will include an approximately equal numbers of elements and the number of the interfacing boundary edges will be minimised. Advance knowledge regarding the number of elements in the final mesh is obtained by training a neural network to predict the number of elements that may be generated for each element from the initial coarse mesh.

The Sub-domain Generation Method (SGM) [2, 60] was originally implemented for planar convex finite element sub-domains using adaptive triangular unstructured meshes. The method comprises the following main components:

- a *Genetic Algorithm*-based optimisation module; and

- a *Neural Network*-based predictive module

The SGM was originally developed in sequential form for use on transputer arrays. Transputer-based systems are typical distributed memory MIMD architectures and usually have a central processor, or ROOT, which has one of its links connected to the HOST system. This processor is therefore responsible for the input-output support of the transputer system. When the idealisation is of a large scale this task may overload the ROOT processor forming a bottle-neck in the analysis procedure. The memory and other performance considerations are detailed in references [2, 60].

The initial mesh is divided into two sub-domains by using a genetic algorithm regulated by an objective function which has a maximum value when the numbers of generated elements are equal in both the sub-domains and the number of interfacing edges is a minimum. This procedure is then applied recursively to each sub-domain.

The coarse initial mesh is thus partitioned into the desired number of sub-domains which are subsequently adaptively re-meshed. This re-meshing may be done concurrently, producing the final distributed mesh for use in a parallel finite element analysis. For generating sub-domain meshes in isolation to one another it is important that compatibility at the interfacing element edges is maintained. This problem may be easily solved on the basis of the technique used in references [2, 60], where the nodal mesh parameter δ_n was used instead of the element mesh parameter δ_e. The nodal mesh parameter δ_n may be readily calculated by nodal averaging.

Sections 5.3.9 and 5.3.10 describe two approaches for the parallelisation of the sequential SGM algorithm. First the neural network predictor is described.

5.3.7 Neural Networks

Artifical Neural Networks (ANNs) stem from neurobiological studies of the brain and the nervous system. In the SGM program, the *backpropagation* (BP) neural networks, as already described in section 5.2.1 are used for approximation purposes. This type of network is based on the backpropagation algorithm [61] and its objective is to optimise a set of connecting weights so that a set of specified input patterns become associated with some output patterns. Once a desired set of weights has been achieved, the trained network may then be tested for some unknown input patterns for which an approximate output is given.

At the start of a training all the weights are set to random values and during the training the network continues adjusting the weights until a desired set of weights has been reached. This means that the training may be stopped once the difference between the desired output values of the training data and the actual output values of the network lies within a satisfactory range. The training may have little difficulty in reaching a desired state, but this may not be true in some cases of BP training where the training is difficult. The difficulty of a network in learning patterns may a result of the composition of the problem in terms of its proper and appropriate mapping for the neural network training or the poor data distribution.

Increasing the number of hidden units and/or increasing the number of hidden layers may contribute to a better learning, however one should bare in mind that too many hidden units or layers may cause the network to memorise the training patterns, (i.e. making perfect match) but be unable to generalize well and give good approximation for unknown patterns. Readers may find more information regarding the problems facing the BP training in references [62] and [63].

The BP neural networks for sub-domain generation

In references [2, 60, 64] a trained BP neural network was used as a predictive module for the decomposition of triangular finite element meshes. The purpose of the trained network is to approximate the number of elements which will be generated within each element of the coarse mesh when adaptive finite element re-meshing is carried out. This information is then used to partition this course mesh into several sub-domains using the genetic algorithm optimisation module.

A similar approach has been carried out in Reference [65] where a BP network was trained by using the data of several finite element meshes but with quadrilateral elements so that the trained network may then produce satisfactory approximations for the number of elements which may be generated in each of the coarse elements after re-meshing. The neural network software used was NETS 2.01 [66] which is based on the classical BP algorithm and is in the public domain.

The training strategy and the network topology:

As it was described in Section 5.3.6 the re-meshing of a coarse background element is done based solely on the basis of the element geometry and the nodal mesh parameters. Considering a mesh of quadrilateral elements only, Figure 5.15 shows the remeshing procedure for a single quadrilateral with all the parameters involved [65]. The element geometry is defined by the four side lengths of the element (s_1, s_2, s_3 and s_4) and the internal angle at the first corner, α. Thus for training purposes a neural network with nine inputs (five for the geometric description and four nodal mesh parameters for each element) and one output (the number of elements generated) was foreseen. (The situation is similar when considering triangular finite elements and is described in references [2, 60].)

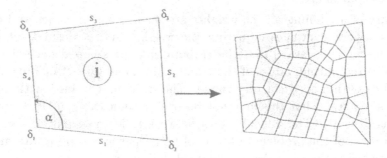

Figure 5.15: The i-th element is considered to be an independent 'sub-domain' for re-meshing

However knowing the fact that each nodal mesh parameter actually represents the size of the fine quadrilateral elements to be generated in the vicinity of the node, the scaling the four sides and the four nodal mesh parameters with one of the nodal mesh parameters would render that mesh parameter to be constant in the data set of the input stimuli. Figure 5.16 shows this new representation of input data with the output number of elements N as opposed to the initial set in Figure 5.15.

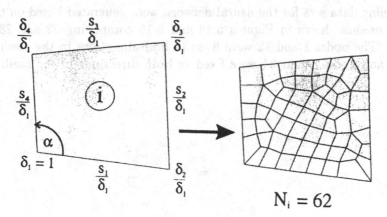

Figure 5.16: Scaled representation of the input data

Hence the neural network has eight inputs and one output. Besides the input and output layers the neural network has four hidden layers in a fully connected scheme as shown in Figure 5.17.

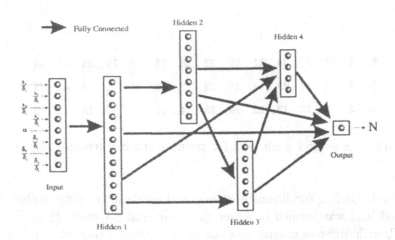

Figure 5.17: The six layer fully connected network used for quadrilateral elements

The training data sets and performance test of the neural network:

Again the basic concept is described with regards to finite element meshes comprising quadrilateral elements. The equivalent description for triangular elements may be found in references [2, 60].

The training data sets for the neural network were generated based on two simple background meshes shown in Figures 5.18 and 5.19 comprising 32 and 28 elements respectively. The nodes 1 and 12 were fixed in both directions in the mesh shown in Figure 5.18 and nodes 1 and 24 were fixed in both directions in the mesh shown in Figure 5.19.

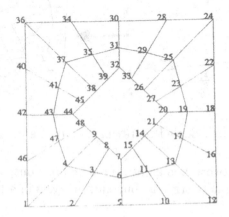

Figure 5.18: The first mesh used for training the NN (comprising 32 elements)

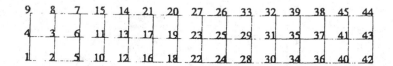

Figure 5.19: The second mesh used for training the NN (comprising 28 elements)

Very simple loading conditions were imposed on these training meshes. One single concentrated load was applied in either the horizontal direction P_x or in the vertical direction P_y with different magnitudes but always one at a time at certain nodes of the idealisation selected randomly. Table 5.2 gives the exact choice of loaded nodes. Thus the first and the second training mesh were subjected to 36 and 22 simple load cases respectively.

For each of these load cases the meshes were analysed, then finite element error analysis was carried out, then the element mesh parameters were determined. By

First training mesh in Figure 5.18			
Load type	Magn.	Loaded nodes	No. load cases
P_x	100	16, 18, 20, 22, 28, 30, 32, 36, 40, 44	10
P_x	200	16, 18, 20, 22, 28, 30, 32, 36, 40, 44	10
P_y	100	4, 5, 7, 19, 20, 28, 30, 32	8
P_y	-100	4, 5, 7, 19, 20, 28, 30, 32	8
		Total number of load cases for this mesh:	36
Second training mesh in Figure 5.19			
Load type	Magn.	Loaded nodes	No. load cases
P_x	100	4, 5, 6, 7, 15, 21, 33, 38, 44	9
P_x	200	4, 5, 6, 7, 15, 21, 33, 38, 44	9
P_y	-100	12, 39, 42, 44	4
		Total number of load cases for this mesh:	22

Table 5.2: The loading cases for the training meshes

nodally averaging these parameters the nodal mesh parameters were calculated and the meshes were adaptively re-meshed using these parameters by the parallel quadrilateral mesh generator [59]. The results of these re-meshings were saved in a file in such a way that it contained the following information for each background element in the training meshes for each load cases: the geometric definition of the element and the nodal mesh parameters in the scaled form described in the previous section (eight values) and the actual number of refined elements generated in that coarse element (one value).

Thus the first training mesh gives $32 \times 36 = 1152$ sets of data. Similarly the second mesh gives $28 \times 22 = 616$ sets of training data. This altogether would result in $1152 + 616 = 1768$ data sets. But some of the data were too close, in particular when the generated number of elements is equal to 1, so these would not affect the training, only increase the computational load. For this reason the training data set was pruned to enhance the training efficiency by removing 254 sets from the data file, leaving 1514 data sets for the training.

The neural network was trained with respect to limiting the *RMS error* values of the training meshes. The RMS error is an overall error measure for a mesh based upon the difference between the predicted (p_k) and the actual (a_k) number of elements generated for each background element:

$$RMS_{error} = \sqrt{\frac{\sum_{k=1}^{N_i}(a_k - p_k)^2}{N_i}} \tag{5.18}$$

where N_i is the number of coarse elements in that mesh.

The network was trained using the parameters in Table 5.3, until the desired accuracy was achieved. Then three new meshes were used to test the performance of the neural network. These meshes and their re-meshed versions are shown in Figures 5.20, 5.21 and 5.22. The RMS error values achieved by the trained network are displayed in Table 5.4.

This network gives enhanced accuracy over the one which was used in the original program of SGM [60] developed for meshes built from triangular finite elements. Its main virtue is the ability to estimate the generation of upto 600 elements per coarse background element with good accuracy.

The training of the neural net could be undertaken in in parallel as described in section 5.2.1 and referecne [16]

Training parameters for the NETS 2.01	
Max. Weight	1.0
Min. Weight	-1.0
Learn Rate	0.25
Momentum	0.50
RMS_{error}	0.0065
No. of Cycles	3000

Table 5.3: The training parameters

Test mesh:	"hook"	"dam"	"ex61"
No. Elems in coarse mesh	18	127	137
No. Elems in refined mesh	909	5390	1999
RMS_{error} values	7.795	4.918	2.789

Table 5.4: The test results of the trained net

Building the neural network into the SGM program

After the training reached the required error tolerance (0.0065 RMS), "flash code" was created by the neural network software. This is a piece of source code written in C which can be built into any other software in a form of a function call. This code contains the final trained values of the weights for the neurons, and applies them to the processing of each new set of input data. It is clear from Sections 5.3.7, 5.3.7 and 5.3.7 describing the topology and the training of the neural network used for sub-domain generation that there are two separate networks for meshes comprising triangular and quadrilateral elements. They are both built into the SGM program as separate C functions and called as neccessary.

In the case of the quadrilateral version this C function has eight input parameters which are the input values for the neural network and a ninth parameter which is the

Figure 5.20: The first mesh ("hook") used for testing the NN (comprising 18 elements) and its re-meshed version (comprising 909 elements)

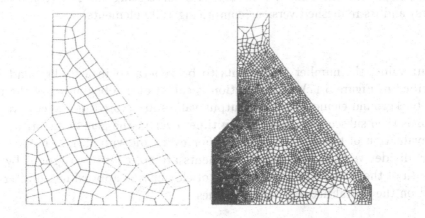

Figure 5.21: The second mesh ("dam") used for testing the NN (comprising 127 elements) and its re-meshed version (comprising 5390 elements)

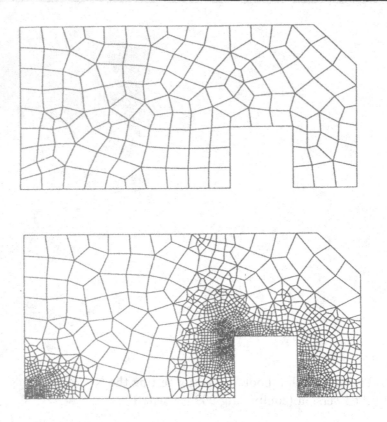

Figure 5.22: The third mesh ("ex61") used for testing the NN (comprising 137 elements) and its re-meshed version (comprising 1999 elements)

output value, the number of elements to be generated in a background element (as described in Figure 5.17). This function is called at the beginning of the program for each background element and the output values are stored in a vector variable. This vector is then subsequently used many times over in the genetic algorithm module for the evaluation of the fitness function. For every individual, i.e. every new position of the divider vector D, the coarse elements are split into two groups by D. Finally for both of the groups the total number of elements is approximated after re-meshing based on the neural network output values:

$$NE_{f1} = \sum_{k=1}^{NE_{i1}} N_k \quad \text{and} \quad NE_{f2} = \sum_{k=1}^{NE_{i2}} N_k \quad (5.19)$$

where NE_{i1} and NE_{i2} are the number of coarse background elements on the first and second parts of the initial mesh after the splitting. The values of NE_{f1} and NE_{f2} are substituted into the fitness function evaluation given later.

5.3.8   Genetic algorithm for mesh partitioning

Cutting the mesh

The SGM was developed for planar convex finite element meshes. The mesh had to be planar and convex because of the algorithm used to bisect the mesh, which works on the following manner: the finite element mesh is cut by a *dividing vector D* in its plane as shown on Figure 5.23. This vector is defined by three variables:

$$D = \{x, y, \theta\}^T \tag{5.20}$$

where:

$$x \in [x_{min}, x_{max}]$$
$$y \in [y_{min}, y_{max}] \tag{5.21}$$
$$\theta \in [0, \pi]$$

The elements of the finite element domain are then divided into two sets by virtue of their centroidal locations, i.e. either to the left or to the right side of the generated vector.

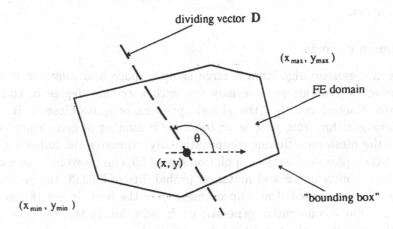

Figure 5.23: Vector D generated to bisect the FE domain

The optimal position of this dividing vector D has to be determined in the optimisation procedure of the domain decomposition accordingly to the conditions described in Section 5.3.6. Thus these three parameters (x, y and θ) become the design variables for the genetic algorithm based optimisation module.

Objective function - fitness function

The optimum criteria is to have an equal number of elements in both the resulting sub-domains whilst the number of interfacing nodes should be at a minimum. This is achieved by an objective function constructed in the following way:

$$g(x, y, \theta) = |NE_{f1} - NE_{f2}| + C_{cf} \tag{5.22}$$

where NE_{f1} and NE_{f2} are the number of finite elements in the first and second half of the mesh respectively, after being cut with the current dividing vector D. Thus if NE_f would represent the number of elements in the full mesh then $NE_f = NE_{f1} + NE_{f2}$. The determination of NE_{f1} and NE_{f2} are described in Section 5.3.7 by equation (5.19). The calculation of the cumulative square root value C_{cf} of the number of interfacing edges C_f in the bisected mesh is detailed for meshes comrising triangular elements in references [2, 60] and for quadrilateral elements in Reference [65].

The genetic algorithm procedure tends to maximise the fitness of its individuals. The function in equation (5.22) is mapped into a *fitness function* $f()$ so that a maximum $f()$ corresponds to a better cut. This is accomplished using the following expression:

$$f(x, y, \theta) = C_{max} - g(x, y, \theta) \tag{5.23}$$

where C_{max} is a suitably chosen large constant number, selected as NE_f in the SGM implementation.

Convergence criteria

By giving the genetic algorithm a large sample space and allowing it to process a large number of generations, one may say with a certain degree of confidence that the solution reached is either the global optimum or quite close to it. To prevent the genetic algorithm from processing too large a number of generations which would render the the mesh-partitioning computationally expensive the following convergence criterion was employed. For the population size of 50, chromosome length of $3 * 10$ bits, crossover probability of 0.8 and mutation probability of 0.0333, the genetic algorithm procedure was terminated if no improvement over the best design (fittest individual) was made within 2 consecutive generations. In addition to this criterion a maximum limit of 500 was set on the number of generations but this criterion was never invoked during the test runs of the SGM.

5.3.9   Recursive bisections on the ROOT with a parallel GA

The first approach to parallelise the SGM method is relatively simple and involves the parallelisation of the genetic algorithm part only. This is the part which is the most time consuming in the algorithm of the SGM. The logical structure of this parallelisation is represented in Figure 5.24.

This algorithm is in reality a task farming [2] parallelisation scheme hence it maintains separate *sub-populations* on each worker processor and they are communicated to and compared on the ROOT. This correspons to an island model as previously discussed. There is no inter-processor communication between these workers. This

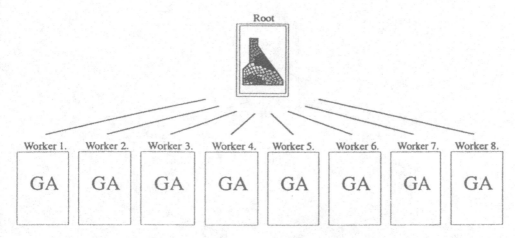

Figure 5.24: Task farming approach to the parallelisation of the SGM

procedure is then called recursively for each sub-domain, until the desired number of splits have been achieved.

On completion of such domain decomposition the parallel mesh generator [2, 58, 59] may be invoked to refine the meshes of each sub-domains. This mesh generator uses the same type of task farming parallelisation scheme.

The disadvantage of this form of parallelisation of the SGM is that although the mesh is decomposed using the coarse mesh, in practice the ROOT processor will have to handle all the information for the refined final mesh. In addition the results of the mesh generation have to be redistributed over the processor network for analysis.

5.3.10   Parallel recursive bisections in a tree structure

In this algorithm each processor uses a local SGM subroutine to decompose the mesh, then sends one of the two resulting sub-domains on to another processor, as shown in Figure 5.25, which will be still idle. After the completion of the sub-domain generation each processor is equipped with part of the initial mesh for adaptive re-meshing and processing during the explicit finite element analysis.

These meshes must be first refined with the adaptive mesh generator, but this mesh generator will generally be sequential or parallel on a single processor. The considerations regarding the newly generated boundary nodes mentioned in Section 5.3.6 still apply with the addition that the lack of global node numbering scheme [2, 67] requires a special strategy for the synchronisation of these nodes. The worker processors must keep track of the constantly changing new positions of their boundary nodes due to the progressing decomposition. This is essential for the parallel finite element analysis routine, which may commence immediately after the completion of the last split on

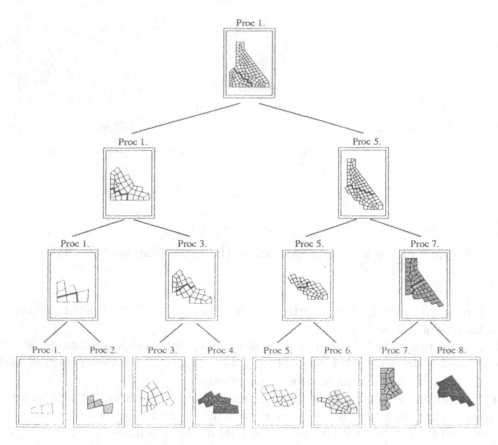

Figure 5.25: Recursive tree structure for the parallelisation of the SGM

each of the processors and the subsequent re-meshing.

Despite these administrative difficulties, this second scheme maintains the initial virtue of the SGM, i.e. to handle the final mesh information in a totally distributed sense. The description of the final refined mesh never goes through the ROOT processor, apart from the final output of the finite element analysis at the very end of the analysis procedure.

Task connectivity scheme

This tree structure approach for the parallelisation of the SGM is slightly more complex than the parallel genetic algorithm technique as it involves a development of a special routing scheme to deliver the description of the partitions to the idle processors as the sub-domain generation proceeds in a logical tree structure. This routing scheme, shown in Figure 5.26, has to conform to the routing algorithm [2, 68] used by the finite element program which will be based on an explicit time-marching finite element algorithm.

The *master task* loads down the coarse finite element mesh description from the HOST, but the analysis is carried out by identical *worker tasks*. During the explicit finite element analysis, these worker tasks have to exchange data relating to the common interfacing finite element nodes at each iteration step. This requires the possibility of direct communication between every worker task because of the arbitrary shape and possible connection layout of the sub-domains. To make this communication as efficient as possible a special routing scheme was developed in references [2, 68] which utilises all the four physical (hardware) links of a transputer, but rendering this hardware specific structure invisible for the application programs.

The structure of this connectivity scheme is best described in relation to Figure 5.26. As it is shown, the components of the application program, i.e. the master task and the worker tasks, have as many communication channels as many worker tasks are in the system. (That is eight in this case.) If one of these tasks is about to communicate with any other it sends a message along the relevant channel from its set, as if the channel was connected to the appropriate task. This is not always physically possible because the transputer has only four physical links and each of them could only be used to carry one (hardware link) channel.

The router task resolves this problem by handling the processor's limited number of physical links as they are connected to each other using these hardware link channels and in doing so it logically creates a virtual channel system permitting general processor to processor communication. An application module has all its channels connected to the appropriate router task, running on the same processor, but these channels are internal 'on-processor' channels. The router task monitors all these channels and on the receipt of the message it adds a very small address header to it. The address headers are used by only the router tasks and passed on to the next router on a neighbouring processor. The router task knows the final destination of the message from the ID number of the internal channel through which it received the message from the application task. When the message finally arrives at the target processor, the router task decodes it (i.e. removes the header) and passes the message to the application program via the appropriate channel.

5.3.11   Example

A domain with the shape of a cross section of a dam, shown in Figure 5.27 with an in-plane horizontal distributed load on the left hand side and the bottom nodes restrained, was uniformly meshed and the initial mesh comprised 127 elements, as shown in Figure 5.28.

The parallel implementation of the SGM was applied to this initial mesh and the mesh was divided into eight sub-domains by performing three recursive bisections, as shown in Figure 5.28. The sub-domains obtained were independently re-meshed

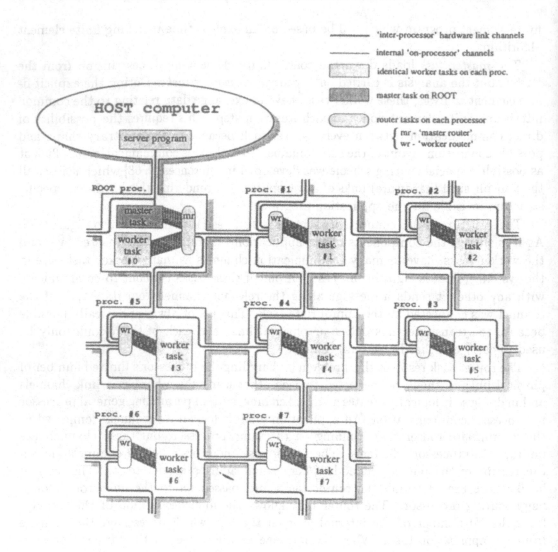

Figure 5.26: Transputer routing scheme for parallel finite element analysis

and the re-meshed sub-domains are shown in Figure 5.29. The resulting *final mesh* comprises 2730 elements.

Here the nodes along the boundaries of the sub-domains were fixed and nodal smoothing was undertaken on the sub-domains distributed among the processors. It would be possible to smooth the nodes on boundaries but this would require inter-processor communication.

Table 5.5 shows the timing results of Parallel SGM for this example. Table 5.6

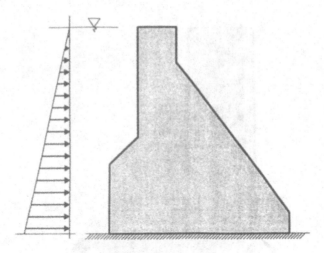

Figure 5.27: SGM Example: domain with a cross sectional shape of a dam

shows the number of elements generated in each sub-domain, and the number of cut interfaces after the re-meshing of these sub-domains.

Timing for:	Rec. 1	Rec. 2	Rec. 3	Total time
Proc. 1.	138.44 s	6.72 s	1.70 s	146.86 s
Proc. 3.			9.80 s	154.96 s
Proc. 5.		77.48 s	10.51 s	226.43 s
Proc. 7.			19.23 s	235.15 s
max. value	138.44 s	77.48 s	19.23.s	235.15 s
	Total execution time:			237.20 s

Table 5.5: PSGM Example: Timing details for the individual splits

Table 5.7 shows the execution times of PSGM on different number of processors and the notional speed-up which can be calculated.

The row "Max. No. of processors used" in Table 5.7 refers to the Parallel version of SGM running on multiple processors. This is the maximum number of processors being used at the end of the procedure, when the progressing tree structure reached its lowest level.

Although the speed-ups for the partitioning are not particularly high the parallel version of the SGM has the inherent advantage that the sub-domains are delivered to the distributed processors without the requirement to form the complete finite element model on the root processor.

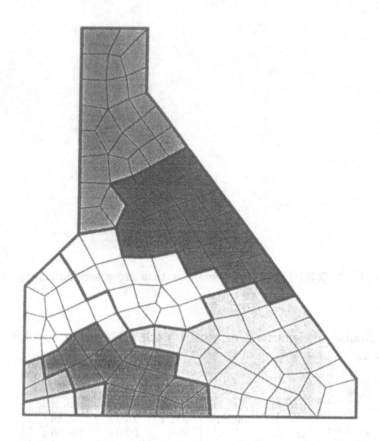

Figure 5.28: SGM Example: the initial mesh (127 elements) divided into eight sub-domains

5.4 Concluding Remarks

In this chapter aspects of: parallel and distributed computing; parallel neural networks; and parallel genetic algorithms have been discussed. The use of these techniques for domain decomposition of unstructured adaptive finite element meshes is described.

Acknowledgements

The research described in this chapter was made possible by the support afforded by HCM Contract No: CHRX-CT93-0390 (DG12COMA) "Advanced Finite Element Solution Techniques on Innovative Computer Architectures".

   The research described in this chapter was supported by Marine Technology Directorate Ltd research contracts: "High Performance Computing for Marine Technology

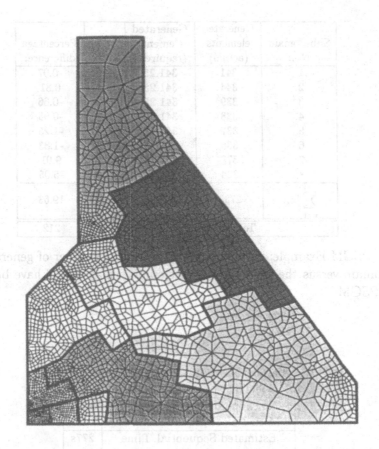

Figure 5.29: SGM Example: the re-meshed sub-domains (2730 elements)

Research", (ref: GR/J22191); and "High Performance Adaptive Finite Element Computations for CAD of Offshore Structures using Parallel and Heterogeneous Systems", (ref: GR/J54017). The research described was also supported by the Systems Architecture Committee of the UK Engineering and Physical Sciences Research Council through the research contract: "Domain Decomposition Methods for Parallel Finite Element Analysis", (ref: GR/J51634).

The authors wish to thank other members of the Structural Engineering Computational Technology (SECT) Research Group in the Department of Mechanical and Chemical Engineering, Heriot-Watt University for useful discussions.

Sub-domain No.:	Generated elements (actual)	Generated elements (required)	Diff.:	Percentage difference:
1.	341	341.25	-0.3	-0.07
2.	344	341.25	2.8	0.81
3.	339	341.25	-2.3	-0.66
4.	338	341.25	-3.3	-0.95
5.	337	341.25	-4.2	-1.25
6.	335	341.25	-6.2	-1.83
7.	372	341.25	30.8	9.01
8.	324	341.25	-17.2	-5.05
$\sum_{i=1}^{n} |x_i|$	2730	2730	67.0	19.63
Total number of cut interfaces:				212

Table 5.6: PSGM Example: Comparison of the actual number of generated elements per sub-domain versus the ideal number of elements that should have been generated using the PSGM

Parallel SGM on a single proc.	277s
Parallel SGM on multiple proc.	237s
No. Sub-domains generated	8
Max. No. of processors used	4
Estimated Sequential Time	277s
Notional Speed-up	1.17

Table 5.7: Comparison of the execution times and the speed-up

References

1. B.H.V. Topping, and J. Sziveri, "Parallel Processing, Neural Networks and Genetic Algorithms", Proceedings of the Second International Conference on Computer Applications Research and Practice, v3, University of Bahrain, April, 1996.

2. B.H.V. Topping and A.I. Khan, *"Parallel Finite Element Computations"*, Saxe-Coburg Publications, Edinburgh, U.K., 1996.

3. A. Geist, A. Beguelin, J. Dongarra, W. Jiang, R. Manchek, V. Sunderam, *"PVM: Parallel Virtual Machine - A User's Guide and Tutorial for Network Parallel Computing"* The MIT Press, Cambrideg, Massachusetts, U.S.A., 1994.

4. W. Gropp, E. Lusk, A. Skjellum., *"Using MPI - Portable Parallel Programming with the Message-Passing System"*, THe MIT Press, Cambridge, Massachusetts, U.S.A., 1995.

5. Donald O. Hebb, *"The Organization of Behavior"*, Wiley Publication Company, New York, U.S.A., 1949.

6. C. C. Jorgensen, *"Neural Network Representation of Sensor Graphs in Autonomous Robot Path Planning"*, IEEE First Int. Conf. on Neural Networks IV, 507-515, San Diego, U.S.A., June 1987.

7. P. Van Der Smagt, B. J. A. Kröse, *"A Real Time Learning Neural Robot Controller"*, in Proceedings of the 1991 Int. Conf. on Artificial Neural Networks, 351-356, Espoo, Finland, June 1991.

8. W. T. Miller III, *"Real Time Application of Neural Networks for Sensor-Based Control of Robot with Vision"*, IEEE Transactions on Systems, Man and Cybernetics 19, 825-831, August, 1989.

9. K. Fukushima, *"Neocognitron: A Hierarchical Neural Network Capable of Visual Pattern Recognition"*, Neural Networks 1, 119-130, 1988,

10. C. Mead, *"Analog VLSI and Neural Systems"*, Addison-Wesley, Reading, MA, U.S.A., 1989.

11. D. Marr, *"Vision"*, W. H. Freeman, San Francisco, U.S.A., 1982.

12. Y. Izui, A. Pentland, *"Speeding Up Back Propagation"* in M. Caudill, (Editor), Theory Track Neural & Cognitive Sciences Track of the Proceedings of the International Joint Conference on Neural Networks, IJCNN-90-WASH-DC, Lawrence Erlbaum Associates, Publishers, New Jersey, United States of America, Vol. 1, 639-642, 1990

13. M. Hagiwara, *"Accelerated Back Propagation using Unlearning based on Hebb Rule"*, in M. Caudill, (Editor), Theory Track Neural & Cognitive Sciences Track of the Proceedings of the International Joint Conference on Neural Networks, IJCNN-90-WASH-DC, Lawrence Erlbaum Associates, Publishers, New Jersey, United States of America, Vol. 1, 617-620, 1990

14. S-B. Cho, J. H. Kim, *"An Accelerated Learning Method with Backpropagation"* in M. Caudill, (Editor), Theory Track Neural & Cognitive Sciences Track of the Proceedings of the International Joint Conference on Neural Networks, IJCNN-90-WASH-DC, Lawrence Erlbaum Associates, Publishers, New Jersey, United States of America, Vol. 1, 605-608, 1990

15. D. E. Rumelhart, G. E. Hinton, R. J. Williams, *"Learning Internal Representation by Error Propagation"*, in D. E. Rumelhart and J. L. McClelland (Eds.) Parallel Distributed Processing: Explorations in the Microstructure of Cognition, Vol. 1 : Foundations., MIT Press, 1986.

16. A. I. Khan, B. H. V. Topping, A. Bahreininejad, *"Parallel Training of Neural Networks for Finite Element Mesh Generation"* in B. H. V. Topping and A. I. Khan (Eds.), Neural Networks and Combinatorial Optimization in Civil and Structural Engineering, 81-94, Civil-Comp Press, Edinburgh, U.K., 1993.

17. Ken L. Parker, Allison L. Thornbrugh, *"Parallelized BP Training and Its Effectiveness"*, in M. Caudill, (Editor), Theory Track Neural & Cognitive Sciences Track of the Proceedings of the International Joint Conference on Neural Networks, IJCNN-90-WASH-DC, Lawrence Erlbaum Associates, Publishers, New Jersey, United States of America, Vol. 2, 179-182, 1990

18. C. Peterson, T. Rognvaldsson, *"An Introduction to Artificial Neural Networks"*, C. Verkerk (Ed.), Proceedings of 1991 CERN School of Computing, CERN Vol. 92, No. 2, 113-170, May 1992.

19. A. Bahreininejad, B.H.V. Topping and A.I. Khan, *"Subdomain Generation Using Multiple Neural Networks Models"*, in B. H. V. Topping, M. Papadrakakis, (Eds.) Parallel and Vector Processing for Structural Mechanics, Civil-Comp Press, Edinburgh, 1994.

20. B.H.V. Topping, A. Bahreininejad, *"Subdomain Generation Using Parallel Q-state Potts Neural Networks Multiple Neural Networks Models"*, in B. H. V. Topping, (Editor) Developments in Neural Networks and Evolutionary Computing for Civil and Structural Engineering, 65-78, Civil-Comp Press, Edinburgh, 1995.

21. J.J. Hopfield, D.W. Tank, *"Neural Computation of Decisions in Optimization Problems"*, Biological Cybernetics, Vol. 52, 141-152, 1985.

22. S. Kirkpatrick, C.D. Gelatt Jr., M.P. Vecchi, *"Optimization by Simulated Annealing"*, Science, 220:4598, 671-680, 1983.

23. F.Y. Wu, *"The Potts Model"*, Reviews of Modern Phys., Vol. 54, No. 1, 235-268, 1982.

24. K.H. Fischer, J.A. Hertz, *"Spin Glasses"*, Cambridge University Press, UK, 1991.

25. J.M. Yeomans, *"Statistical Mechanics of Phase Transition"*, Oxford University Press, U.S.A., 1992.

26. J. Hertz, A. Krogh and R.G. Palmer, *"Introduction to the Theory of Neural Computing"*, Addison-Wesley Publication Company, U.S.A., 1991.

27. G. Bilbro, R. Mann, T.K. Miller III, W.E. Snyder, D. E. Van den Bout, M. White, *"Optimization by Mean Field Annealing"*, David S. Touretzky (Ed), Advances in Neural Information Processing Systems 1, Morgan Kaufmann Pub. Inc., 91-98, 1989.

28. C. Peterson, J.R. Anderson, *"Neural Networks and NP-complete Optimization Problems; A Performance Study on the Graph Bisection Problem"*, Complex Systems, Vol. 2, No. 1, 59-89, 1988.

29. C. Peterson, B. Söderberg, *"A New Method for Mapping Optimization Problems onto Neural Networks"*, Int. J. Neural Syst., Vol. 1, No. 1, 3-22, 1989.

30. C. Peterson, J. R. Anderson, *"A Mean Field Learning Algorithm for Neural Networks"*, Complex Systems, Vol. 1, 995-1019, 1987.

31. A.I. Khan, B.H.V. Topping, *"Parallel adaptive mesh generation"*, Computing Systems in Engineering, vol. 2, No. 1, 75-102, 1991.

32. A.I. Khan, B.H.V. Topping, *"Subdomain Generation for Parallel Finite Element Analysis"*, Computing Systems in Engineering, Vol. 4, Nos. 4-6, 473-488, 1993.

33. A. Bertoni, M. Dorigo, *"Implicit Parallelism in Genetic Algorithms"*, Artificial Intelligence, V 61, N 2, 307-314, 1993.

34. J.J. Grefenstette, J.E. Baker, *"How Genetic Algorithms Work: A Critical Look at Implicit Parallelism"*, Proceedings of the Third International Conference on Genetic Algorithms, Morgan Kaufmann Publishers, Inc, 70-79, Virginia, U.S.A., 1989.

35. Mühlenbein, H, Schleuter, M G- and Krämer, O (1988) *"Evolution Algorithms in Combinatorial Optimization"*, Parallel Computing, V 7, N 1, 65-88, 1988.

36. Pettey, C C, Leuze, M R and Grefenstette, J J (1987) *"A Parallel Genetic Algorithm"*, Genetic Algorithms and Their Applications: Proceedings of the Second International Conference on Genetic Algorithms, Lawrence Erlbaum Associates Publishers, 155-161, Cambridge, MA, U.S.A., 1987.

37. Tanese, R (1987) *"Parallel Genetic Algorithm for a Hypercube"*, Genetic Algorithms and Their Applications: Proceedings of the Second International Conference on Genetic Algorithms, Lawrence Erlbaum Associates Publishers, pp 177-183, Cambridge, MA, U.S.A., 1987.

38. Collins, R J and Jefferson, D J (1991) *"Selection in Massively Parallel Genetic Algorithms"*, Proceedings of the Fourth International Conference on Genetic Algorithms, Morgan Kaufmann Publishers, Inc, pp 249-256, San Diego, CA, U.S.A., 1991.

39. Manderick, B and Spiessens, P (1989) *"Fine-Grained Parallel Genetic Algorithms"*, Proceedings of the Third International Conference on Genetic Algorithms, Morgan Kaufmann Publishers, Inc, pp 428-433, Virginia, U.S.A., 1989.

40. Schleuter, M G- (1992) *"Comparison of Local Mating Strategies in Massively Parallel Genetic Algorithms"*, Proceedings of Parallel Problem Solving by Nature, 2, R Männer and B Manderick (Editors), Elsevier Science Publishers B V, pp 65-74, Brussels, Belgium, 1992.

41. Grefenstette, J J (1981) *"Parallel Adaptive Algorithms for Function Optimization"*, Technical Report N CS-81-19, Vanderbilt University, Computer Science Department, Nashville, U.S.A., 1981.

42. Leite, J P B and Topping, B H V (1995) *"Improved Genetic Operators for Structural Engineering Optimization"*, CIVIL-COMP95, Proceedings of the Sixth International Conference on Civil and Structural Engineering Computing: Developments in Computational Techniques for Structural Engineering, in Developments in Neural Networks and Evolutionary Computing for Civil and Structural Engineering, Civil-Comp Press, 151-165, Cambridge, UK., 1995

43. T.C. Fogarty, Huang, R (1990) *"Implementing the Genetic Algorithms on Transputer Based Parallel Processing Systems"*, Proceedings of Parallel Problem Solving by Nature, H-P Schwefel & R Männer (Editors), Springer-Verlag Publishers, 145-149, Dortmund, FRG, 1990.

44. P. Neuhaus, *"Solving the Mapping-Problem - Experiences with a Genetic Algorithm"*, Proceedings of Parallel Problem Solving by Nature, H-P Schwefel & R Männer (Editors), Springer-Verlag Publishers, 170-175, Dortmund, FRG., 1990.

45. S. Wright, *"Stochastic Process in Evolution"*, Stochastic Models in Medicine and Biology, J Gurland, Ed, University of Wisconsin Press, 199-241, U.S.A., 1964.

46. R. Shonkwiler, F. Mendivil, A. Deliu, *"Genetic Algorithms for 1-D Fractal Inverse Problem"*, Proceedings of the Fourth International Conference on Genetic Algorithms, Morgan Kaufmann Publishers, Inc, 495-501, San Diego, CA, U.S.A., 1991.

47. R. Shonkwiler, *"Parallel Genetic Algorithms"*, Proceedings of the Fifth International Conference on Genetic Algorithms, Morgan Kaufmann Publishers, Inc, 495-501, Illinois, U.S.A., 1993.

48. B.H.V. Topping, J.P.B. Leite, *"Parallel Genetic Models for Structural Optimization"*, Submitted to publication, 1996.

49. N. Mansour, G.C. Fox, *"A Hybrid Genetic Algorithm for Task Allocation in Multicomputers"*, Proceedings of the Fourth International Conference on Genetic Al-

gorithms, Morgan Kaufmann Publishers, Inc, 466-473, San Diego, CA, U.S.A., 1991.

50. H. Mühlenbein, M. Schomisch, J. Born, *"The Parallel Genetic Algorithm as Function Optimizer"*, Proceedings of the Fourth International Conference on Genetic Algorithms, Morgan Kaufmann Publishers, Inc, 271-278, Virginia, U.S.A., 1991.

51. E.-G. Talbi, P. Brassière, *"A Parallel Genetic Algorithm for the Graph Partitioning Problem"*, ACM International Conference on Supercomputing, Cologne, Germany, 1991

52. P. Jog, J.Y. Suh, D.V. Gucht, *"Parallel Genetic Algorithms Applied to the Travelling Salesman Problem"*, SIAM Journal of Optimization, V 1, N 4, 515-529, 1991

53. S.-C. Lin, W.F. Punch, E.D. Goodman, *"Coarse-Grain Parallel Genetic Algorithms: Categorization and New Approach"*, Accepted for publication, Parallel and Distributed Processing, Dallas, TX, U.S.A., 1994.

54. W.F. Punch, R.C. Averill, E.D. Goodman, S.-C. Lin, Y. Ding, Y.C. Yip, *"Optimal Design of Laminated Composite Structures using Coarse-Grain Parallel Genetic Algorithms"*, Computing Systems in Engineering, V 5, N 4-6, 415-423, 1994.

55. H. Adeli and N-T. Cheng, *"Concurrent Genetic Algorithms for Optimization of Large Structures"*, ASCE - Journal of Aerospace Engineering, V 7, N 3, 276-296, 1994.

56. H. Adeli and S. Kummar, *"Distributed Genetic Algorithm for Structural Optimization"*, ASCE - Journal of Aerospace Engineering, V 8, N 3, 156-163, 1995.

57. H. Adeli and S. Kummar, *"Concurrent Structural Optimization on Massively Parallel Supercomputer"*, ASCE - Journal of Structural Engineering, V 121, N 11, 1588-1597, 1995.

58. A.I. Khan and B.H.V. Topping, *"Parallel Adaptive Mesh Generation"*, Computing Systems in Engineering, Vol. 2, No 1, pp. 75-102, Pergamon Press, UK, 1991.

59. B.H.V. Topping and B. Cheng, *"Parallel Adaptive Quadrilateral Mesh Generation"*, in Advances in Computational Structures Technology, Civil-Comp Press, Edinburgh, UK, 1996.

60. A.I. Khan and B.H.V. Topping, *"Sub-domain Generation for Parallel Finite Element Analysis"*, Computing Systems in Engineering, Vol. 4, Nos 4-6, pp. 473-488, Pergamon Press, UK, 1993.

61. D. E. Rumelhart, G. E. Hinton and R. J. Williams, *"Learning Internal Representation by Error Propagation"*, in D.E. Rumelhart and J. L. McClelland (Eds.)

Parallel Distributed Processing: Explorations in the Microstructure of Cognition, Vol. 1: Foundations., MIT Press, U.S.A., 1986.

62. A.I. Khan, B.H.V Topping and A. Bahreininejad, " *Parallel Training of Neural Networks for Finite Element Mesh Generation*", Neural Networks and Combinatorial Optimization in Civil and Structural Engineering, 81-94, Civil-Comp Press, Edinburgh, UK, 1993.

63. B.H.V. Topping and A. Bahreininejad, *"Neural Computing for Sturctural Mechanics"* Saxe-Coburg Publications, Edinburgh, UK, 1996.

64. B.H.V. Topping and J. Sziveri, *"Parallel Sub-domain Generation Method"*, Developments in Computational Techniques for Structural Engineering, 449-457, Civil-Comp Press, Edinburgh, UK, 1995.

65. J. Sziveri, B. Cheng, A. Bahreininejad, J. Cai, G. Thierauf and B.H.V. Topping, *"Parallel Quadrilateral Subdomain Generation"*, in Advances in Computational Structures Technology, Civil-Comp Press, Edinburgh, UK, 1996.

66. P. T. Baffes, *Nets User's Guide, version 2.01*, NASA, Lyndon B. Johnson Space Center, U.S.A., 1989.

67. B.H.V. Topping and A.I. Khan, *"Parallel Computation Schemes for Dynamic Relaxation"*, Engineering Computations, Vol. 11, 513-548, Pineridge Press Ltd, Swansea, UK, 1994

68. A.I. Khan and B.H.V. Topping, *"A Transputer Routing Algorithm for Non-linear or Dynamic Finite Element Analysis"*, Engineering Computations, Vol. 11, 549-564, Pineridge Press Ltd, Swansea, UK, 1994

CHAPTER 6

NEURAL NETWORKS AND FUZZY LOGIC
IN ACTIVE CONTROL OF MECHANICAL SYSTEMS

P. Venini
University of Pavia, Pavia, Italy

ABSTRACT

A few applications of neural networks and fuzzy logic in active control of systems are presented. Rigid and flexible, linear and nonlinear, stable and unstable structures are investigated. The basics of neural networks are not covered unlike the essentials about fuzzy reasoning that are highlighted. Within the presented control algorithms neural networks accomplish different tasks. They cover cases in which the control action is computed according to a neural-only paradigm all the way through simpler applications where the role of neural networks is to replicate the behavior of conventional controllers. Nonlinear civil structures under seismic excitation and nonlinear rigid systems as an inverted pendulum and a container-ship are among the systems studied in much detail. Attention is focused on the physics of the problem, on the objectives of the neurocontroller as well as on its practical implementation. Numerical simulations and experimental tests are illustrated to validate the presented approaches.

6.1. Introductory remarks

Aim of this chapter is to present and discuss a variety of applications of neural networks in the broad area of active control of systems and structures. The objects of the investigation may be roughly listed as follows:

1. NN-based algorithms for the active control of linear and nonlinear structural systems. At this regard, we will examine in some detail
   - constitutively nonlinear systems, e.g. elasto–plastic and endochronic MDOF oscillators;
   - geometrically nonlinear systems, e.g. the inverted pendulum problem and a rigid–body idealization of a container ship;
   - time–delay nonlinearities induced by sensors and processors;
   - unmodelled nonlinearities such as friction between cart wheels and rack in the inverted pendulum experiment;
2. use of neural networks as approximators of complex nonlinear mappings. We will examine the case of a nonlinear oscillator controlled by a receding–horizon technique. Such control strategy requires so many on–line computations that make it inapplicable for real–time applications. The way out to this problem is shown to be a neural network that is trained off–line to approximate the receding–horizon mapping between sensored response and control action.

Not many details will be conversely given about the theoretical issues pertaining to the neural networks with which the reader is assumed to be familiar, see also chapter 1. An important remark at this purpose is that in what follows only backpropagation networks are used and so is to be intended the phrase "neural network" throughout the text. Several technical papers from the recent literature are cited within the text but the pioneering book [22] is worth a remark of its own as an (even now) up–to–date contribution in the area of active control with neural networks.

6.1.1. Motivations

All the problems listed before and investigated next have been object of analysis for long time by means of conventional strategies. Most of all, the active control of linear systems finds its roots in the mathematical, aerospace and electronic literatures long before the development of NN–based approaches. The adoption of the neural paradigm is then advised by a few reasons including:

   - the emerging consciousness that active control is a viable alternative for the design of civil structural systems, such as supertall buildings and long–span bridges. For such kind of structures, having several thousands of degrees-of–freedom, analytical (classical) techniques may be prohibitive whereas NN-based strategies are likely to be more viable;
   - NN-based controllers do not require the mathematical model of the system under control;
   - NN-based controllers are inherently nonlinear and seem therefore to be well-suited for strongly nonlinear systems as those mentioned above;

- the control action computed by a trained network derives from a massively distributed process that makes the network fairly sensitive to local failures;
- the capability of learning from examples makes the network adaptive, i.e. new training cases may be generated to adapt the neural controller to modifications that have interested the structure as those due to damage induced by strong external agencies.

There are of course also negative features that characterize neural schemes, and namely:

- the stability of the controlled (closed–loop) system has to be checked "by hands", e.g. by extensive numerical simulations, since we still lack analytical proofs of stability as those based on Lyapunov stability theory;
- as we will see, the training procedure is usually quite lengthy, definitely more cumbersome, for example, than the solution of the Riccati matrix equation which represents the core of many optimal control strategies.

6.1.2. Overview

Section 6.2 gives the basic ideas about neurocontrol and presents a few numerical applications to nonlinear MDOF systems. The (complex) procedure that leads to the development of the neurocontroller is therein described and commented in detail, leaving to subsequent sections the extensions needed to handle geometric nonlinearities in a correct way. The model problem of the shear–type structure under seismic excitation is examined. Hysteretic restoring forces are introduced by means of the Bouc–Wen model [3]. Section 6.3 deals with the well–known problem of the stabilization of an inverted pendulum in upright position. This is quite peculiar with respect to most of the current literature that is usually concerned with flexible stable system. Leaving to the (near) future the development of a neural–only controller, we propose herein the adoption of a mixed scheme that couples a conventional LQ controller with a neural compensator. The tremendous advantages of the proposed technique over the uncompensated strategy are described by means of numerical as well as experimental studies. Section 6.4 introduces the basics of receding–horizon (RH) control [27, 17] for nonlinear systems. The RH controller requires the solution of a nonlinear programming problem at each sampling–time and is therefore not applicable, at least directly. A neural approximation of the RH controller based on off-line training is therefore presented as an appealing alternative for allowing real–time applications of such powerful strategy. Finally, elements of fuzzy reasoning are outlined in section 6.5 [35] and then applied to the control of the manouvres of a containership [40].

6.2. NEURAL CONTROL OF NONLINEAR HYSTERETIC STRUCTURES

This section introduces a nonlinear model of civil structure subject to random seismic excitation to which a few NN–based active control algorithms are applied. These control strategies are described in detail and tested numerically on the MDOF hysteretic structures discussed next.

6.2.1. The Bouc–Wen model

One of the most widely used models for nonlinear components in earthquake engineering is the Bouc–Wen model [3] that allows to include hysteretic phenomena into the analysis within a rather simple framework. Referring to shear–beam idealizations, the i–th interstory restoring force F_{s_i} reads

$$F_{s_i} = \alpha_i k_i x_i + (1 - \alpha_i) k_i D_{y_i} v_i, \tag{6.1}$$

where α_i is the post–to–pre yielding stiffness ratio, k_i is the initial elastic stiffness, x_i is the relative displacement between adjacent storeys and D_{y_i} is the yielding displacement. The new hysteretic variable v_i is a state variable so that each story of the building is now described by a triplet of state variables, i.e. displacement, velocity and hysteretic variable. The nonlinear constitutive law is then introduced by means of the nonlinear equation

$$\dot{v}_i = D_{y_i}^{-1} \{ A_i \dot{x}_i - \beta_i |\dot{x}_i| |v_i|^{n_i - 1} v_i - \gamma_i \dot{x}_i |v_i|^{n_i} \}, \tag{6.2}$$

where A_i, β_i and γ_i control the shape of the hysteresis loop and n_i determines the smoothness of the loop and the degree of nonlinearity. In particular, as n_i approaches infinity, an elastic perfectly plastic behavior is attained. Equations (6.1) and (6.2) define completely the Bouc–Wen endochronic model that need to be coupled to the classical second–order equation of motion. One should notice that, unlike standard plasticity models that call for an incremental formulation, the adoption of the Bouc–Wen model allows to use finite quantities likewise the linear case. This allows the usage of classical methods of integration for nonlinear differential equations avoiding expensive techniques based on Newton–Raphson iterative schemes.

6.2.2. Global equations of motion

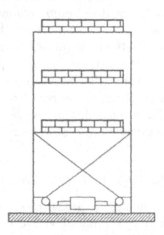

FIGURE 6.1. Typical 3DOF shear-type structure with active tendons

Let us focus our attention on a n degree–of–freedom system which is excited by a ground base acceleration $\ddot{x}_g(t)$ and controlled by a control vector $u(t)$ having dimension r, see Figure 6.1. The goal of this study is of course to train a proper network able to issue the "correct" control signal as an answer to the measurements made available on–line by a few sensors placed on the structure. The vector equation of motion of the system may be written as

$$M\ddot{x}(t) + C\dot{x}(t) + K_e x(t) + K_h v(t) = Hu(t) + \Gamma\ddot{x}_g(t), \tag{6.3}$$

where $M = n \times n$ mass matrix, $C = n \times n$ damping matrix, $K_e = n \times n$ initial stiffness matrix, $K_h = n \times n$ hysteretic stiffness matrix, $H = n \times r$ topological matrix indicating the location of the actuators, $\Gamma = n \times 1$ vector distributing the inertial forces due to the base motion to each story. A n-dimensional vector f is then introduced that groups all the equations of type 6.2, one for each story of the structure, i.e.

$$\dot{v} = f(\dot{x}, v), \quad f_i(\dot{x}_i, v_i) = D_{y_i}^{-1}\{A_i\dot{x}_i - \beta_i|\dot{x}_i||v_i|^{n_i-1}v_i - \gamma_i\dot{x}_i|v_i|^{n_i}\}, i = 1,\ldots,n. \tag{6.4}$$

Toward the adoption of classical time integration schemes for first–order vector differential equations, we introduce a $3n$ state vector $z(t)$, a $3n \times r$ matrix B and a $3n \times 1$ vector W as

$$z(t) = \left\{\begin{array}{c} x \\ v \\ \dot{x} \end{array}\right\}, \quad B = \left[\begin{array}{c} 0 \\ 0 \\ M^{-1}H \end{array}\right], \quad W = \left\{\begin{array}{c} 0 \\ 0 \\ M^{-1}\Gamma \end{array}\right\}. \tag{6.5}$$

One may then write the nonlinear state–space equation of motion as

$$\dot{z} = g[z(t)] + Bu + W\ddot{x}_g(t), \tag{6.6}$$

where $g[z(t)]$ is a $3n$ vector, depending nonlinearly on the state $z(t)$, that reads

$$g[z(t)] = \left\{\begin{array}{c} \dot{x} \\ f(\dot{x}, v) \\ -M^{-1}[C\dot{x} + K_e x + K_h v] \end{array}\right\}. \tag{6.7}$$

There exist several numerical schemes for the integration of (6.6) that were used in the numerical simulations to come. The fourth–order Runge–Kutta method, for instance, amounts to writing the solution vector $z(t)$ as

$$z(t) = z(t - \Delta t) + \frac{1}{6}\Delta t\,[A_0 + 2A_1 + 2A_2 + A_3], \tag{6.8}$$

in which Δt is the integration time step and A_i, $i = 1,\ldots,4$ are $3n$ time–dependent vectors that read

$$A_0 = \{g[z(t - \Delta t)] + Bu(t - \Delta t) + W\ddot{x}_g(t - \Delta t)\},$$

$$A_1 = \left\{g[z(t - \frac{1}{2}\Delta t) + \frac{1}{2}\Delta tA_0] + Bu(t - \frac{1}{2}\Delta t) + W\ddot{x}_g(t - \frac{1}{2}\Delta t)\right\},$$

$$A_2 = \left\{g[z(t - \frac{1}{2}\Delta t) + \frac{1}{2}\Delta tA_1] + Bu(t - \frac{1}{2}\Delta t) + W\ddot{x}_g(t - \frac{1}{2}\Delta t)\right\},$$

$$A_3 = \{g[z(t) + \Delta tA_2] + Bu(t) + W\ddot{x}_g(t)\}.$$

6.2.3. Neurocontrol schemes

6.2.3.1. *General remarks*

The overall idea behind neurocontrol is to train a neural network to be used on line that is able to issue a control signal based on some inputs, typically some components of the response measured by appropriate sensors. When a supervised training is adopted, as is always the case when using backpropagation networks as is done herein, the key point is to decide which kind of mapping should the neurocontroller approximate. Ever since introductory studies where conducted in the early nineties, [11, 25, 26, 42], it was soon realized that three are the possible philosophies at the base of neural–based control schemes as listed below.

1. A neurocontroller is trained to approximate the input–output mapping from another controller. As correctly stated in [13] this approach is not fully satisfactory since the neurocontroller cannot then behave any better than the original controller. However, as shown in [44], there exist control strategies that are theoretically powerful but practically of no use because of the cumbersome on–line computations required. The approximation of such mappings by a neural network, see also [27], is therefore essential for the practical implementation of some control methods and is not only a matter of taste or fashion;

2. a neural network approximating the inverse dynamics of the structure is the second choice at disposal. The state of the structure at current time and at a few past sample times represents the input to the network whose output is the control signal to the structure. The idea is that the control action for a given state should be equal to minus the action that would have caused that state. Such a control action is expected to anneal the dynamics of the system. This is surely true for a completely controllable system and holds only in an approximate sense in general [11, 25]. A demonstration of the effectiveness of this control strategy will be given in the numerical examples at the end of this section;

3. an emulator network is trained first with the aid of which the neurocontroller is in turn trained. The emulator network approximates the direct (nonlinear) system dynamics and allows to compute the sensitivity of the control signal with respect to the structure response. Peculiar and crucial in this approach is the need to backpropagate the control error through the emulator network with the goal of finding the input that produces a desired output, whereas usual backpropagation is done with the goal of adjusting weights, i.e. the internal structure of the neural network. Different methodologies were developed on this purpose, [4, 7, 11, 13], that will be described next.

In what follows, the above approaches numbered 2 and 3 are described in some detail and numerically tested on a MDOF nonlinear frame, whereas the receding horizon case, i.e. no. 1 above, is left to forecoming sections.

6.2.4. Neurocontrol by inverse dynamics network

This method requires the training of a single neural network that will serve as controller [25, 42]. The idea is to let the neural network learn the inverse of the transfer function between the actuator(s) and the measurable output of the system. Training the network and details about its usage as a controller are objects the of the forecoming sections.

6.2.4.1. *Neural network training*

The first step is the selection of the architecture of the neural network. On the one hand, theorems exist that ensure the capability of backpropagation networks with (at–least) two hidden layers to approximate any nonlinear mapping within a desired accuracy [6, 14]. On the other hand, no general guidelines are available to choose the architecture but the experience of the analyst. Usually, and herein also, a trial–and–error procedure is adopted until a satisfactory network design is found. For completeness sake, the adaptive method developed in [2] is to be cited where one starts with a small number of neurons per layer and add new neurons whenever the global error of approximation does not decrease for a fixed number of iterations. The main advantage of this approach is the capability of escaping from local minima, well known problem common to all non–convex optimization techniques. The number of neurons of the input layer depends on how many sensors are available to monitor the state of the structure. Let then m denote such number. To allow the neural network a time–dependent view of the response, which is crucial when the response is nonlinear, each measured quantity should be held for a certain number of sampling periods, say n_s, so that the size of the input layer may be determined as $n_i = m \times n_s$. The output layer has as many nodes, say n_c, as there are actuators and frequently $n_c = 1$. What one should decide then is the number of hidden layers and the number of neurons per layer. A frequent choice is to have two hidden layers whose number is determined either by trial–and–error [11, 42] or adaptively [2, 26]. Once the (initial) architecture is chosen, a sufficient number of training epochs, i.e. input–output pairs with which the network is to be presented, must be generated. These epochs must be representative of the mapping one wants the network to approximate. We have used a swept–sine technique as the one adopted in classic system identification that works as follows. Let $I_\omega = \{\omega_a, \omega_b\}$ be the frequency range of interest, i.e. one wants the neural controller to be most effective in the presence of an excitation whose predominant spectral component belongs to I_ω. Then, N artificial signals are generated as

$$f_n(t) = \sum_{j=1}^{n_n} a_j \sin(\omega_j + \varphi_j), \ \omega_j \in I_\omega, \ n = 1, \ldots, N, \tag{6.9}$$

where $\|a_j\|_{\ell^2} = \sum_{j=1}^{n_n} a_j^2$ may vary from signal to signal to account for different earthquake intensities, n_n is the number of waves of each signal and the φ_j's are random phase shifts. One should notice that when n_n is large this amounts to the well–known technique proposed in [34] for the simulation of (white) stochastic processes. For each

$f_n(t)$, the system dynamics is simulated in the time interval $[0, M\Delta t]$ where Δt is the time step for numerical integration and M is the total number of sample times. If one denotes by $y_n(t)$ the global response corresponding to $f_n(t)$, the k–th input/output pair to be learned by the neural network may be written as

$$i_k = \{y_{t_k}\ y_{t_{k-1}}\ \cdots y_{t_{k-n_s}}\},$$
$$o_k = f_n(t_k).$$

Figure 6.2 shows the procedure for training the inverse dynamics neurocontroller. A

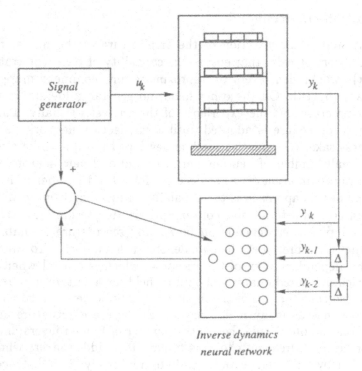

FIGURE 6.2. Training the inverse–dynamics neurocontroller

few more issues about the procedure are the following ones:

- measurement delay, say τ, may be included in the procedure by a simple shift in the output pattern that becomes $o_k = f_n(t_k + l \times \Delta t)$ where l is a proper integer such that $\tau = l \times \Delta t$. This way the network is taught to approximate the inverse dynamics of the system making at the same time a prediction over a time–lag equal to the estimated delay induced by measurements and computing time. The delay τ does not conversely include actuator dynamics that may be explicitly incorporated in the equations of motion [11, 13, 25];
- the input to the neural network may also contain the excitation when the latter is measurable;
- a random disturbance should always be added to guarantee a realistic setting, also in view of experimental applications, see sections 6.3–6.3.3.

6.2.4.2. *Neural controller in operation*

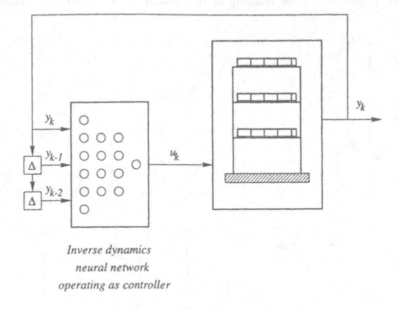

Inverse dynamics
neural network
operating as controller

FIGURE 6.3. Inverse–dynamics neurocontroller in operation

Figure 6.3 shows the usage of the trained network as a neurocontroller which is of course reversed with respect to the training phase. Given the immediate past history of the structural response, the network computes the control action to be acted upon the structure. In synthesis, one may conclude that the philosophy behind this approach is rather simple: a neural network that retains the inverse dynamics of the structure, i.e. the inverse of the transfer function between the control command and the response, is able to generate a sequence of control forces that are able, in principle, to anneal the response of the structure.

6.2.5. Inverse neural controller aided by an emulator network

Although successful applications of the above strategy are reported in the literature [25, 42], it carries a limitation that may be described as follows. When the neurocontroller is operating properly, the components of the response with which the network is fed–back are reasonably small so that subsequent control commands issued by the neurocontroller tend to be small as well. However, in the presence of significant excitations, the structural response will increase again until the neurocontroller takes over reducing the response. This originates time windows in which the controller does its job correctly alternate to others in which the effectiveness is drastically reduced. A remedy to this problem is to separate the global response into a contribution due to the control action and another due to the external disturbances. Then one has to feed the network with the part of the response that is caused by the external excitation only. This may be computed as the difference between the global response and that due to the control action only that is in turn evaluated by means of an emulator neural network, see Figure 6.4. The training of the emulator is left to next sections where

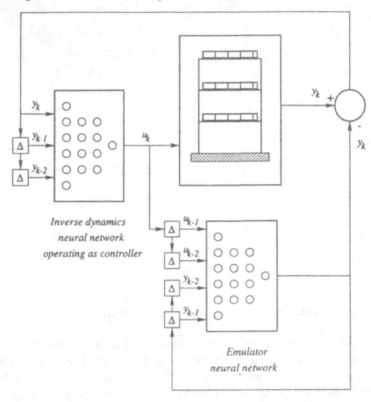

FIGURE 6.4. Inverse–dynamics neurocontroller with the aid of an emulator

a neural network emulating the dynamics of the system is used off–line to train the neurocontroller.

6.2.6. Neurocontrol with the aid of an emulator

6.2.6.1. *Introduction*

The approach to be introduced now translates in quantitative terms the qualitative reasoning behind any control strategy. Roughly speaking such reasoning is as follows: given the current state of the structure and having in mind a desired state, ideally the undamaged structure at rest, what is the control command to be issued that drives the structure "most" toward the desired state? Inherent in the question is the call for a relationship between the *error* that is computed in term of divergence of the *structure state* from rest and the *action* that is conversely available as an *actuator force*. It is shown hereafter that a scheme using a neurocontroller aided by an emulator network represents an excellent tool for computing this transfer from the structure error to the control command.

6.2.6.2. *Training the emulator*

The emulator neural–network may be trained (off–line) as using the procedure of Figure 6.5. The immediate past histories of the response signal $\{y_{k-1} \ldots y_{k-h}\}$ and of the input signal $\{u_k u_{k-h}\}$ are the inputs to the network whose output is the current state $y_k i$ to be contrasted with the actual one computed by integrating the equations of motion. The error may then be evaluated and a standard backpropagation applied to adjust the weights until convergence is reached.

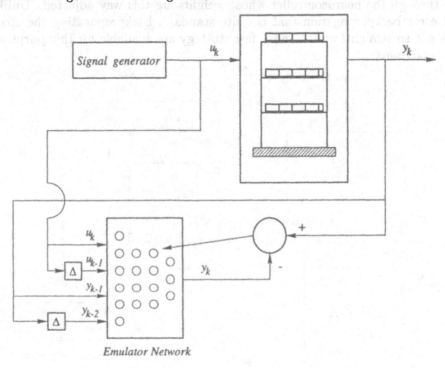

FIGURE 6.5. Training of the emulator neural network

6.2.6.3. *Training the neurocontroller*

As mentioned in section 6.2.6.1, the most difficult issue in the development of a (neuro) controller is that the error is at the structure level whereas the action is at the actuator level. If backpropagation neural networks (BPNN) are used, this obstacle translates into an immediately clear computational problem. In fact classical training of BPNN's requires to compute the error at the output layer of the net and back-propagate it through all the layers, adjusting internal weights according to the Delta rule or some faster variations. In the case at hand, see Figure 6.6, the output of the untrained neurocontroller is a trial control action that should be compared with the "correct action" to make the output error available. However, the "correct action" is not at disposal and such is therefore the output error. What is conversely available, see again Figure 6.6, is the error that comes from a comparison, usually a simple difference, between the actual state of the system and a desired one as computable by the control criterion box. The role of the emulator network is exactly to provide the mapping between the structure state error and the control error, so as to allow a proper training of the neurocontroller. The trained emulator made available with the procedure explained in section 6.2.6.2 is then used to backpropagate the structure error, with fixed weights, that allows to compute the control error to be in turn backpropagated through the neurocontroller whose weights are this way adjusted. Unlike the control error backpropagation that is quite standard, backpropagating the structure error is not so straightforward and a few strategy are available for this purpose that are discussed next.

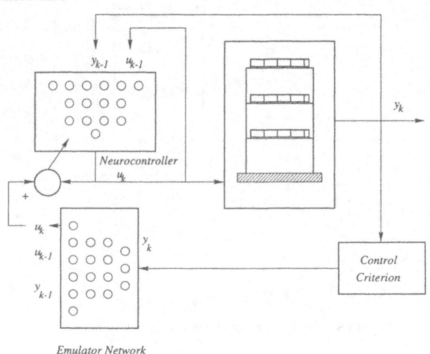

FIGURE 6.6. Training of neurocontroller

6.2.6.4. *Backpropagation through the emulator*

We have to solve the following inverse problem. Given is the architecture of the emulator network whose mapping may be concisely written as

$$\mathcal{E}: \mathcal{R}^N \to \mathcal{R}^M, \quad y = \mathcal{E}(x), \tag{6.10}$$

where, within the numerical studies to follow, $N = 8$, $M = 1$ and

$$x = [\ddot{x}_1(t - 4\Delta t) \; \ddot{x}_1(t - 3\Delta t) \; \ddot{x}_1(t - 2\Delta t) \; \ddot{x}_1(t - \Delta t) \; u(t - 3\Delta t) \; u(t - 2\Delta t) \; u(t - \Delta t) \; u(t)], \tag{6.11}$$

$$y = [\ddot{x}_1(t)]. \tag{6.12}$$

Given now the output vector y, i.e. the structure state error, and seven out of eight components of the input vector, i.e. the four accelerations of the first floor and the last three control actions, the objective is to compute the inverse of $\mathcal{E}$ with respect to $u(t)$ only, i.e. to find the control action that meets the requirement of making the current acceleration equal to the desired one. A first approach to solve the problem is the one presented in [7] that uses a single–hidden layer network but the method may be easily generalized. Let the error function be

$$E_c = \frac{1}{2}\|y_d - y\|_{\ell^2}^2,\qquad (6.13)$$

where y_d is the desired output. By the chain rule, the change of the weights W_{ij}^C in the controller network may be then computed as

$$\Delta W_{ij}^C = -\eta \frac{\partial E_c}{\partial W_{ij}^C} = -\eta \sum_h \frac{\partial E_c}{\partial u_h}\frac{\partial u_h}{\partial W_{ij}^C},\qquad (6.14)$$

where η is the learning rate and u_h is the h–th component of the control vector. It is then clear that the evaluation of ΔW_{ij}^C calls for a backpropagation of E_c through the emulator network with no weight changes to compute the control error, followed by a second backpropagation of the control error through the neurocontroller to adjust the relevant weights. To compute Δu consider first the h–th neuron in the hidden layer of the emulator network. The variation of its output H_h may be written as

$$\Delta H_h = -\eta \frac{\partial E_c}{\partial H_h} = -\eta \sum_o \Delta_o(K+1)W_{oh}^E,\qquad (6.15)$$

where

$$\Delta_o = \frac{\partial F(Net_o)}{\partial Net_o}\left[y_d^o(k+1) - y^o(k+1)\right],\qquad (6.16)$$

and W_{oh}^E is the weight connecting node o in the output layer and node h in the hidden layer of the emulator network and Net_o is the input to the node o in the output layer from the hidden layer. The variation in the control action may then be computed as

$$\Delta u_i = -\eta \frac{\partial H_h}{\partial u_i} = -\eta \frac{\partial F(Net_h)}{\partial Net_h}\sum_h W_{hi}^E \Delta H_h,\qquad (6.17)$$

where W_{hi}^E is the weight connecting node h in the hidden layer and node i in the input layer and Net_h is the input to the node h in the hidden layer from the input layer. Latter equations make the control error available so that the neurocontroller may be trained for the entire duration of the earthquake episode. In addition to the above semi–analytical scheme for backpropagating the output error through the emulator, a more direct approach is at disposal [4, 11] where small increments are given to the current control command in a finite–step fashion until either the control criterion is satisfied or a saturation of the actuator takes place.

6.2.6.5. *The control criterion*

The control criterion, see Figure 6.6, incorporates in some sense the overall philosophy of the control method. Standard L^1, L^2 or L^∞ norms may be used to measure the discrepancy between the current and the target state and the choice between these alternatives depends of course on whether the main concern is on the energy or peak reduction or so on. Another fundamental issue is to decide whether the criterion should be based on the present state only or should look a few steps ahead. In the former case, the criterion may lead to too an authoritative controller if the requested response reduction is too large. In order to smooth the action of the controller one may either fuzzify the input–output relationship of the neurocontroller [13] or use a control criterion of predictive type whose objective is to mitigate the average response over a finite horizon of time [4].

6.2.7. Numerical Studies

A three degree–of–freedom system is analyzed made of three identical storeys whose properties are given in Table 6.1 and 6.2. The control scheme used in this case is the last of the ones discussed above, i.e. a neurocontroller is trained with the aid of an emulator and then used on line with no further networks. As to the emulator,

TABLE 6.1. Physical properties of each of the three storeys

Mass m [tons]	Damping c [KN sec/m]	Stiffness K_e [KN/m]	Stiffness ratio
10.000	692	600.000	0.2

TABLE 6.2. Parameters of the Bouc–Wen model

A	β	γ	n	D [m]
1.0	0.5	0.5	5.0	0.04

Figure 6.7 presents a comparison between the actual response as computed by means of the Runge–Kutta integration scheme and the one predicted by the emulator network. The disturbance signal is a white noise simulated with the technique of Shinozuka [34]. The difference between the two signals is so small to be hardly visible at the two time scales reported in the Figure. The controller was then tested in the presence of the El Centro seismic record that is reported in Figure 6.8 along with the consequent control command. To judge about the effectiveness of the control, one should first recall that the output error was solely based on the acceleration of the first floor which is therefore the one to be checked primarily. Figure 6.9 allows some insight showing the reduction in the first floor displacement and acceleration. One may see that in both the uncontrolled and controlled cases there is a drift that is not recovered, at least in the four–second window object of analysis but the uncontrolled one is definitely tolerable.

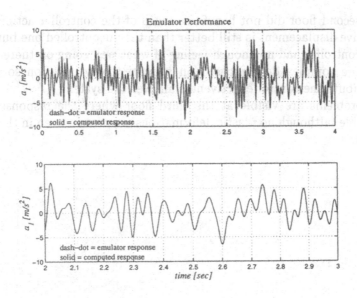

FIGURE 6.7. Testing the emulator neural network

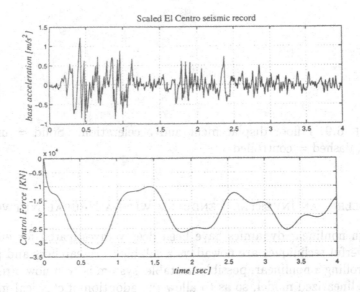

FIGURE 6.8. El Centro seismic record and corresponding control action

Even more informations may be gathered by looking at Figure 6.10 that shows all the hysteresis cycles in the uncontrolled and controlled cases. Two comments are in order

- the controller has made its job so nicely that the first floor has remained in the linear range. This is in excellent agreement with the requirement we expected from the controller;

- the second floor did not benefit so much of the controller action. Its residual relative displacement is still better than the uncontrolled one but it seems like the controller had no enough feeling of what was going on there. This advises a more complex design based on a multi–storey criterion so as to equally distribute the control effectiveness all over the system;
- no problems are visible at the third floor where the response reduction is sensible (although no plastic deformations are present even in the uncontrolled case).

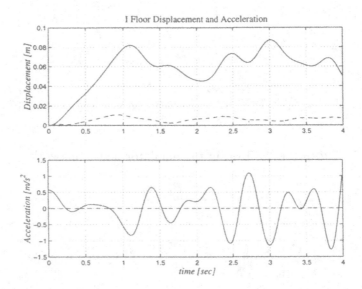

FIGURE 6.9. I floor displacement and acceleration. Solid = uncontrolled, dashed = controlled

6.3. CONTROLLING AN INVERTED PENDULUM WITH A NEURAL NETWORK

Although nonlinear dynamics have been deeply investigated for at least thirty years and powerful techniques are nowadays available to researchers and practitioners [23, 24], controlling a nonlinear, possibly unstable, system is even now often attempted recurring to a linearized model, so as to allow the adoption of classical methods from linear system and control theory [16]. As long as system nonlinearities are small, i.e. when the deviation from some equilibrium position is negligible, such an approach works properly. In this case the potentially nonlinear system does not fully exhibit its nonlinear nature and linear control strategies are able to stabilize and regulate the system. However, imperfections and disturbances are likely to force the system away from the small displacements region toward a truly nonlinear behavior. If this happens, the linearized model is no longer a good descriptor of the system dynamics and the computed controller, based on the linearized system, is therefore unable to operate

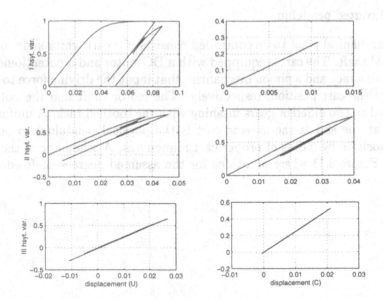

FIGURE 6.10. Hysteresis cycles for the three floors. Left (U)ncontrolled - Right (C)ontrolled

correctly.

Motivated by the foregoing considerations, the use of a neural, inherently nonlinear, compensator is herein suggested. The objective is to let the LQ regulator operate on an equivalent undisturbed system, the effects of disturbances being taken care of by the neural network. The design of such neural compensator requires the definition of the tasks to be accomplished, the consequent way to train the network using experimental data, and numerical as well as experimental on–line validation of the trained compensator. All these issues are investigated in the sections to come.

Coming to the content of this part, section 6.3.1 introduces notations, derives the equations of motion and computes the eigenvalues of the open–loop system showing its instability. Section 6.3.2 shows a detailed derivation of the steps needed to design the LQ controller with particular emphasis on theoretical issues. Section 6.3.3 is devoted to the experimental results and comments on the performance of the derived LQ controller. Limitations of the LQ approach are highlighted in Section 6.3.4 where motivations are given that advise a combined usage of the LQ controller and a neural compensator. The compensator is based on the concept developed in Section 6.3.5 and designed according to the guidelines given in Section 6.3.6. Section 6.3.7 presents the results of the experimental verification of the compensation scheme. The compensated system exhibits excellent stability properties and very small response amplitudes as opposed to the LQ controlled uncompensated system that may become unstable under severe disturbances.

6.3.1. The inverted pendulum

The mechanical plant to be controlled consists of a cart which slides on a ground stainless steel shaft. The cart is equipped with a DC motor and a potentiometer. These are coupled to a rack and a pinion mechanism that input the driving force to the system and measure the cart position, respectively. The motor shaft and the potentiometer are connected to two different gears meshing with the toothed rack. A uniform massive rod hinged at one end at the moving cart is the physical pendulum to be stabilized in upright position by means of proper cart manouvres. A schematic of the pendulum is shown in Figure 6.11 where notations for the assumed degrees–of–freedom are also introduced.

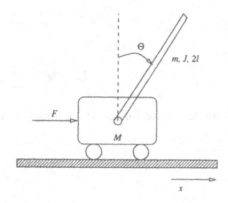

FIGURE 6.11. Schematic of the pendulum and notations

As to the physical properties of the system, M is the mass of the cart that is modeled as a SDOF massive point, m, J and 2ℓ are the mass, baricentric moment of inertia and length of the rod, respectively. Table 6.3 presents the value of these quantities for the system under study.

TABLE 6.3. Physical properties of the experimental model

M [kg]	m [kg]	J [kg m²]	2ℓ [m]
0.455	0.210	6.5×10^{-3}	0.305

The Lagrangian coordinates of the system are x, the position of the cart on the rack, and θ, the angle of rotation of the rod with respect to the vertical direction. If one neglects the effects of friction and denoting by F the action of the DC motor on the cart wheel, the equations of motion may be written as

$$(M + m)\ddot{x} + m\ell\ddot{\theta}\cos\theta - m\ell\dot{\theta}^2\sin\theta = F, \qquad (6.18)$$

$$\frac{4}{3}m\ell^2\ddot{\theta} + M\ell\ddot{x}\cos\theta - Mg\ell\sin\theta = 0. \qquad (6.19)$$

One should notice that friction effects do not appear in (6.18) and (6.19). In fact, modeling friction is a very hard task [1], especially when toothed wheels are in contact. Instead of trying to derive a more realistic analytical model, that would however be extremely difficult to be used for on line integration, we design a neural compensator of friction and other disturbances that does not require any analytical model. The state vector is now introduced as $y = \{y_1\ y_2\ y_3\ y_4\}' = \{x\ \theta\ \dot{x}\ \dot{\theta}\}'$ along with the nonlinear two-dimensional mapping f that may be derived from (6.18) and (6.19) and results in the state–space motion equation as

$$\dot{y} = f(y, F). \tag{6.20}$$

As is well known, the equilibrium positions of the systems are of type

$$\begin{aligned}
y_1 &= \tilde{y}, \\
y_2 &= k\pi,\ k\ \text{integer}, \\
y_3 &= 0, \\
y_4 &= 0,
\end{aligned} \tag{6.21}$$

where $\tilde{y}$ is any position of the cart on the rack and odd values of k correspond to stable positions whereas even values of k give unstable positions. For the time being, the objective is to design a controller based on the linearized dynamics of the system in the vicinity of any unstable equilibrium position. The state equation linearized about $y = \{0\ 0\ 0\ 0\}^T$ reads

$$\dot{y} = Ay + BF, \tag{6.22}$$

where

$$A = \begin{bmatrix} 0 & I \\ -M^{-1}S & -M^{-1}C \end{bmatrix}, \tag{6.23}$$

and

$$M = \begin{bmatrix} M+m & m\ell \\ M\ell & \frac{4}{3}m\ell^2 \end{bmatrix},\quad S = \begin{bmatrix} 0 & 0 \\ 0 & -Mg\ell \end{bmatrix},\quad B = \left\{ \begin{array}{c} 0 \\ 0 \\ M^{-1}\left\{ \begin{array}{c} 1 \\ 0 \end{array} \right\} \end{array} \right\}. \tag{6.24}$$

In (6.23), C is a 2×2 damping matrix that is neglected herein and 0 and I are 2×2 null and identity matrices, respectively. As well known and physically evident, the system is open–loop unstable. The square roots of its eigenvalues are 0, 0, corresponding to allowed rigid body modes, and

$$\omega = \pm\sqrt{\frac{g(M+m)}{\frac{4}{3}\left(m + \frac{m^2}{M}\right)\ell - m\ell}}. \tag{6.25}$$

A complete description of the system requires to specify how the force F acting upon the cart is computed and applied. The force F is generated by the DC motor coupled with the toothed rack by a pinion and is given by the equation

$$F = \alpha_1 u + \alpha_2 \dot{x}, \tag{6.26}$$

where u is the voltage to the motor, $\dot{x}$ is the speed of the cart and α_1 and α_2 are constants. Using (6.22) and (6.26), one may incorporate the term $\alpha_2\dot{x}$ into the state matrix and rewrite the state equation of the system having the voltage u as input. The linearized continuous time model of the system finally reads

$$\dot{y} = A_c y + B_c u. \tag{6.27}$$

The discrete–time state–space realization of the system is then computed with a sampling time $T = 0.005$ seconds for usage within all the experimental tests reported hereafter. One may therefore write

$$y(k+1) = A_D y(k) + B_D u(k) \tag{6.28}$$

where the matrices A_D and B_D were computed using standard continuous–to–discrete conversion algorithms, e.g. as implemented in the MATLAB software environment [20].

6.3.2. Synthesis of the LQ controller

The objective of this section is to design a digital controller capable of stabilizing the system, i.e. able to keep the rod in vertical position. The controller is also expected to regulate the position of the cart along the track. In synthesis, the controller must stabilize the closed loop system while moving the cart toward any desired position. Formally, given the discrete–time linear system of (6.28), the goal is to find a control law

$$u(k) = g[y(k)], \tag{6.29}$$

so that the system described by (6.28) and (6.29) is stable. Hereafter we use the well–known LQ control technique, the basics of which are also outlined.

6.3.2.1. *LQ control problem*

Given the quadratic cost function

$$J = \frac{1}{2} \sum_{k=1}^{N} \left[(y(k) - y_{ref})^T Q (y(k) - y_{ref}) + u(k)^T R u(k) \right], \tag{6.30}$$

where y_{ref} is the time–independent reference state to be tracked, and Q and R are positive definite weighting matrices, the objective is to find the sequence $u(k)$ that minimizes J. Basic results from LQ control theory include [32]:

- for $N = \infty$, the optimal control law is given as

$$u(k) = K(y(k) - y_{ref}) \tag{6.31}$$

where K is the optimal feedback gain matrix obtained by solving an algebraic Riccati matrix equation;

- under weak conditions on A_D, B_D and Q, one may show that the feedback law of (6.31) stabilizes the closed loop system that is then described by the state–space equation

$$y(k+1) = (A_D + B_D K)y(k) - B_D K y_{ref}; \qquad (6.32)$$

- minimizing the quadratic cost J not only forces the plant output to follow the reference but also forces the control variable u to be close to zero in performing its job. The value of the gain K is optimal only with respect to the chosen matrices Q and R. The real design problem that arises when applying this technique is, in fact, the choices of Q and R which are often determined iteratively.

6.3.2.2. *Application to the pendulum*

Choosing the weighting matrices Q and R is typically an iterative procedure. Roughly speaking the ratio between the norms of Q and R is governed by the authority that the controller should have and by the energy that may be supplied in operation. Furthermore, the single entries in Q weigh the relative importance between different state variables and their correlation, if any. For the problem at hand, the rotation of the pendulum is by far the most important variable since it is the one that governs the possible instability of the system. This aspect should of course be reflected in the choice of Q_{22}. Any choice of Q and R must then be checked a–posteriori at least by looking at the closed loop poles as well as by checking that the computed gains are compatible with the power made available on line. After some trials, the following choice was made

$$Q = \begin{bmatrix} 0.25 & 0 & 0 & 0 \\ 0 & 2 & 0 & 0 \\ 0 & 0 & 0.04 & 0 \\ 0 & 0 & 0 & 0.04 \end{bmatrix}, \quad R = 10^{-4}, \qquad (6.33)$$

resulting in the optimal gain

$$K = \{27.7 \ 225.5 \ 41.2 \ 24.09\}. \qquad (6.34)$$

All the consequent closed–loop discrete–time eigenvalues lie inside the unit circle and are given as 0.2982, 0.9197, $0.9955 \pm 0.0038i$, thus ensuring stability of the regulated system.

6.3.3. Experimental results

The experimental setup with electrical and electronic components is sketched in Figure 6.12. Two potentiometers measure the position of the cart x and the rod angle θ and pass the measures to the computer via the electronic board. The control program computes the control law on line and finds the value of the voltage that is to be applied to the DC motor by the power circuit. Available measurements do not include the cart velocity $\dot{x}$ and the rod angular velocity $\dot{\theta}$ but are needed by the LQ controller that

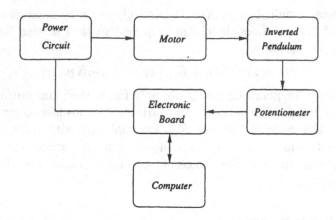

FIGURE 6.12. Schematic of the hardware of the system

calls for a complete state information. A Kalman filter as well as a simpler backward difference scheme were used and tested experimentally to serve this purpose.

The block diagram of the controller is shown in Figure 6.13 where the presence of two low–pass first–order filters is to be noted. They are used to reduce high-frequency noise on the measurements. Filter parameters a and b are chosen to obtain a cutoff

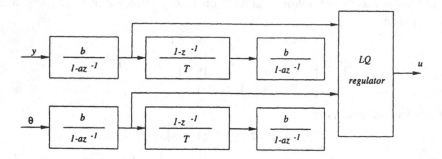

FIGURE 6.13. Control block diagram

frequency of 10 hertz, T being the sampling. The LQ controller block performs the weighted summation of the state variable values using the entries of K as weights.

Given the above description of the system a few successful experimental results are hereafter presented regarding the LQ regulated uncompensated system. In fact, if one recalls that the final strategy we propose is a combined use of an LQ regulator and a neural compensator, it is mandatory to show first that the LQ regulator has good stabilizing properties when operating on a slightly disturbed system. Then, the neural compensator will be introduced to take care of the most severe disturbances.

6.3.3.1. *Equilibrium perturbation experiment*

In the first experiment one wants to keep the cart in a desired position, e.g. the middle point of the rack, and the rod in upright position. This has to be done in the face of a perturbation to the rod that was realized "by hand", by simply hitting the free end of the rod. Two such perturbations are shown in Figure 6.14 where one may see the ability of the control to accomplish its task. Notice that a saturation takes place at five volts meaning that the perturbation intensity was quite challenging for the control system that was nevertheless able to stabilize the pendulum.

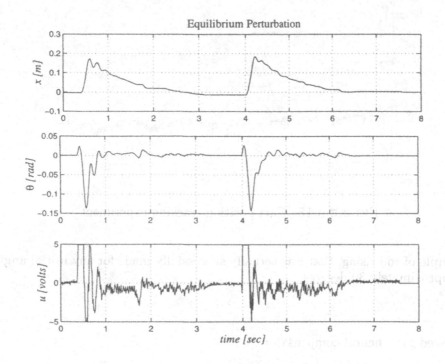

FIGURE 6.14. Perturbation experiment

6.3.3.2. *Cart positioning experiment*

This experiment aims at moving the cart to any given position on the rack. In Figure 6.15 the setpoint was chosen as $y_{1_{ref}} = x_{ref} = 0.3\ m$. One may see that the goal is achieved with a reasonably limited control action as well as with tolerable rod rotations.

6.3.3.3. *Rod rising experiment*

The last experiment shows the ability of the controller to force the system to the equilibrium position starting from a perturbed initial condition. Figure 6.16 illustrates

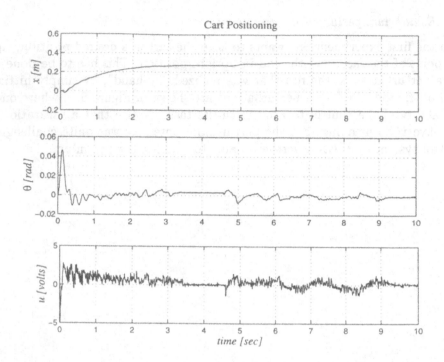

FIGURE 6.15. Cart position tracking experiment

an example of rod rising, that was actually successfully made for any initial angle less than approximately 30 degrees.

6.3.4. Need for a neural compensator

The inverted pendulum control problem has been treated in previous sections and solved according to the following procedure: a nonlinear model of the real plant has been derived and the LQ technique applied to the linearized model. Linear system theory says that the stability of a linear system does not depend on input signals. Therefore, it might be concluded that once the stability of the LQ regulated pendulum is proven, then every input disturbance results in a stable behavior. There are, however, at least two reasons that suggest to search for disturbance compensation techniques.

On the one hand, the real inverted pendulum deviates from linearity due to non-linear geometric effects as well as saturation effects and friction forces that are quite difficult to model and even more to account for via on–line computations. On the other hand, we shall see later that there exist disturbances for which the LQ regulated pendulum fails to stabilize the pendulum.

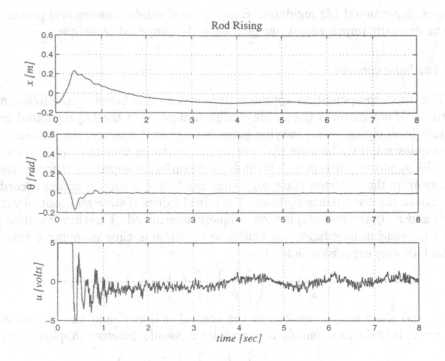

FIGURE 6.16. Rod rising experiment

The presence of strong nonlinearities and the lack of a reliable mathematical model are typical situations in which classical control strategies may be hardly applicable while neural networks may represent the right tool to approach the problem. Next sections give an analytical/experimental proof of the capability of a properly trained neural network to avoid instability of a nonlinear system while drastically reducing the response amplitude. The disturbed pendulum is governed by the state equation

$$\dot{y} = h(y, H), \tag{6.35}$$

that differs from (6.20) only in that H comprises both the control action F and the disturbances, e.g. due to friction effects. One then writes $H(t) = F(t) + F_d(t)$ and (6.35) becomes

$$\dot{y}(t) = h(y(t), F(t) + F_d(t)). \tag{6.36}$$

As proved previously, the linearized system may be stabilized thanks to a control law of the form

$$u(t) = K[y(t) - y_{ref}] \tag{6.37}$$

computable with the LQ technique. Unfortunately, such controller is not robust so that input disturbances as well as large initial conditions are likely to degrade the performance and/or induce instability in the absence of a proper compensation. The following sections present a control scheme in which a neural network provides an estimation of the disturbance F_d, ready to be used for compensation purposes in parallel to a more conventional LQ regulator. Experimental validations are also provided that show the dramatic improvement one gets with the proposed technique.

6.3.5. The basic concept

The direct compensation approach proposed next is based on the following considerations. If one disturbs the stable equilibrium point of the LQ regulated inverted pendulum by applying a time varying extra voltage to the motor, the system deviates from its quiescent state because the controller tries to balance the rod by moving the cart in the opportune direction. Therefore, a disturbance signal $F_d(t)$ forces the state $y(t)$ away from the reference state y_{ref}, thus producing a control action according to (6.37). Given the disturbance $F_d(t)$ and the initial values of state and control variables, i.e. $y(0)$ and $F(0)$, the flow $y(t)$ is then uniquely determined. Therefore, switching from state to Lagrangian coordinates, a nonlinear continuous–time mapping q exists such that the following expression holds true:

$$\left\{ \begin{array}{c} \ddot{x} \\ \ddot{\theta} \end{array} \right\} = q \left(\left\{ \begin{array}{c} \dot{x} \\ \dot{\theta} \end{array} \right\}, \left\{ \begin{array}{c} x \\ \theta \end{array} \right\}, U, d \right), \tag{6.38}$$

where U and d are scaled versions of the control action F and the disturbance F_d, respectively. It is then reasonable to search for a pseudo–inverse mapping of type

$$d = q^{-1} \left(\left\{ \begin{array}{c} \ddot{x} \\ \ddot{\theta} \end{array} \right\}, \left\{ \begin{array}{c} \dot{x} \\ \dot{\theta} \end{array} \right\}, \left\{ \begin{array}{c} x \\ \theta \end{array} \right\}, U \right), \tag{6.39}$$

that, without recurring to accelerations, may be approximated in the discrete–time as

$$d(k) \approx q^{-1} \left(\left\{ \begin{array}{c} \dot{x}(k) \\ \dot{\theta}(k) \end{array} \right\}, \left\{ \begin{array}{c} \dot{x}(k-1) \\ \dot{\theta}(k-1) \end{array} \right\}, \left\{ \begin{array}{c} x(k) \\ \theta(k) \end{array} \right\}, U(k) \right). \tag{6.40}$$

The problem is now how to compute the nonlinear mapping q. If the mathematical model was completely known, than it would be possible to use it to find q analytically and get d by inversion. However, when significant uncertainty is present and important aspects neglected, as friction and saturation effects in our case, q must be found in an approximate way using experimental data.

6.3.6. The compensation scheme

The block diagram of the regulated and compensated system is shown in Figure 6.17 where both the LQ controller and the neural compensator are present. One may see that the total input voltage applied to the system results from three contributions, namely the effect of the *LQ* controller, that depends exclusively on the discrepancy between reference and measured states, the effect of the neural compensator and that of the disturbance. The goal is to let the network take care of the disturbance part, so that one ends up with an equivalent, disturbance–free system, that may be successfully controlled by the LQ regulator. The neural network block computes the mapping

$$U_{NN}(k) = \mathrm{NN}_w(s(k)), \tag{6.41}$$

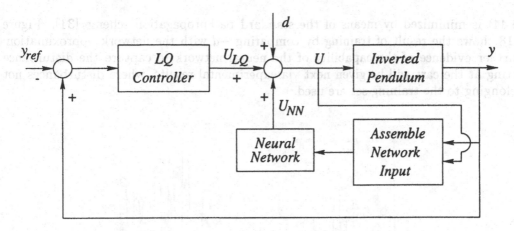

FIGURE 6.17. Compensated system in operation

where NN_w denotes a backpropagation network with given weights stored in the vector w and

$$s(k) = \left\{ x(k), \dot{x}(k), \dot{x}(k-1), \theta(k), \dot{\theta}(k), \dot{\theta}(k-1), U(k) \right\}, \qquad (6.42)$$

is the input vector to the neural network. The block "Assemble Network Input" extracts from the state history the correct value of the network input. The objective of training a neural network to approximate the disturbance $d(k)$ is pursued and detailed in the next section.

6.3.6.1. *Training the neural compensator*

Any friction force between cart and rack, external perturbation and unexpected DC motor behavior may be considered a disturbance to be compensated by the neural network. The latter should be capable of performing the estimation

$$NN_w(s(k)) \approx -d(k). \qquad (6.43)$$

Training patterns may be successfully generated by exciting the regulated pendulum with a known sequence of disturbances $d(k)$ and storing the generated sequence of response vectors $s(k)$, that comprise control actions $U(k)$ as well. The optimization problem having the weight vector w as unknown may be written as

Find w that minimizes the square error $E = \dfrac{1}{2} \displaystyle\sum_{k=1}^{N_k} ||NN_w(S(k)) + d(k)||^2,$ (6.44)

where N_k is the total number of sampling times considered, and is actually a variable itself to be chosen by the analyst. A network with two hidden layers was used. The first layer contains ten neurons with sigmoidal activations while the second layer is a purely linear activation function neuron, i.e. a simple multiplier. The disturbance d is chosen as a filtered random piecewise constant signal ranging in amplitude from -3.5 to 3.5 [Volts], jumps occurring every 100 time steps. The quadratic error function in

(6.44) is minimized by means of the standard backpropagation scheme [31]. Figure 6.18 shows the result of training by comparing $-d$ with the network approximation. Further evidence of the capability of the neural network to capture the disturbance acting at the cart will be given next via experimental results, where disturbances not belonging to the training set are used.

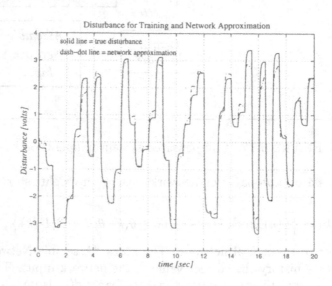

FIGURE 6.18. Validating the trained neural compensator

6.3.7. Experimental results

Experiments were then conducted at the Process Control Laboratory of the University of Pavia to validate the neural compensation scheme developed above. Two cases are reported next. In both cases the reference state is given as $\{x\ \theta\ \dot{x}\ \dot{\theta}\} = \{0\ 0\ 0\ 0\}$, i.e. quiescent conditions with the cart in the initial position and the rod in upright position.

In the first case the disturbance is modeled as a continuous periodic signal with peak value of 4 Volts, see Figure 6.21. It is expected that the LQ controller alone might handle the effect of such "regular" disturbance, but the question is whether the stable motion that results is acceptable as far as the oscillation amplitude is concerned. We can gain insight thanks to Figures 6.19 and 6.20 that show the cart displacement $x(t)$ and the rod rotation $\theta(t)$ in the uncompensated (dash-dot line) and compensated (solid line) case. As expected, smooth periodic disturbances do not cause severe stability problems and the LQ controller does its job nicely. However, it is apparent that the neural compensator was able to remove most of the disturbance effects thus alleviating the burden for the LQ controller that practically operates on an undisturbed system.

Table 6.4 shows the maximum values for cart displacement and rod rotation. The neural network compensator has reduced the response by about 1000 %. Figure 6.21 shows the external disturbance (dash–dot line) and its approximation done by the neural network. It is quite evident that the trained network has learned to extract from the system dynamics the disturbance that acts at the cart.

TABLE 6.4. Uncompensated vs. compensated maxima of y and θ

Case	min y [m]	max y [m]	min θ [rad]	max θ [rad]
Uncompensated	-0.175	0.120	-0.130	0.150
Compensated	-0.020	0.009	-0.015	0.015

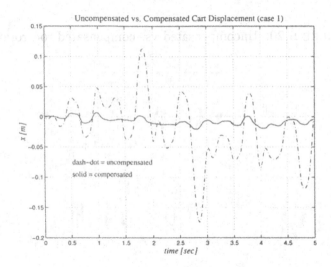

FIGURE 6.19. Uncompensated vs. compensated cart displacement

In the second experiment the disturbance is modeled as a piecewise constant function, changing its value every 200 time steps, see Figure 6.24. Such an irregular disturbance, characterized by jumps of different amplitudes is considered as a benchmark for showing the limitations of the LQ regulator alone and checking the capability of the neural compensator to stabilize the system. In fact, experimental evidence, see Figures 6.22 and 6.23, shows that the system stability is lost because of such disturbance. Uncompensated responses $x(t)$ and $\theta(t)$ cannot be followed experimentally since instability occurs. It is clear that the disturbance causes large rotations that make the linearized system inapplicable for control purposes. The LQ controller has to operate on a system that is too far from the one it was designed for. Therefore, the closed loop system is no longer stable and the pendulum falls down. Conversely, when the neural compensator is present, the effect of the disturbance is practically annealed and the

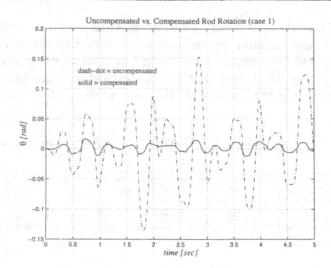

FIGURE 6.20. Uncompensated vs. compensated rod rotation

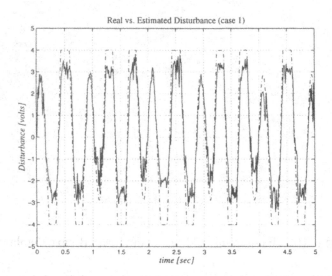

FIGURE 6.21. Real vs. estimated disturbance

stability margin is quite good. Figure 6.24 shows the approximation that the network has made on line of the disturbance that is quite acceptable as well.

6.3.8. Concluding remarks

A neural compensation technique for use on open–loop unstable nonlinear systems has been presented. The combined use of the neural compensator with a linear regulator results in remarkable improvements in terms of stability with respect to the

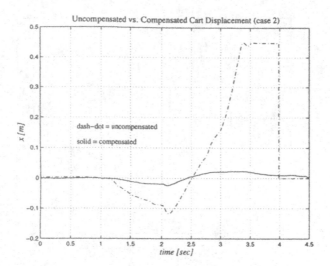

FIGURE 6.22. Uncompensated vs. compensated cart displacement

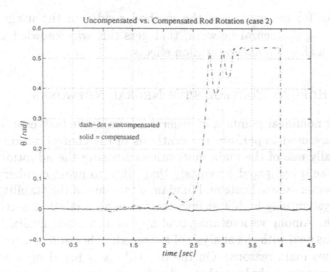

FIGURE 6.23. Uncompensated vs. compensated rod rotation

regulated system without compensation.

Numerical and experimental results were reported showing that:

1. the stability margin of the compensated system is considerably larger than the one of the regulated but not compensated system;

2. when the uncompensated system is stable as well, there is anyway a tremendous reduction in terms of response amplitude due to the action of the compensator.

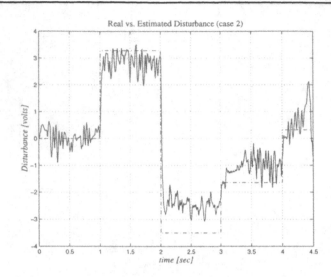

FIGURE 6.24. Real vs. estimated disturbance

A key–point for the success of the method resides in the usage of experimental data for training the neural network, that gets this way knowledge of unmodeled phenomena such as friction and saturation effects.

6.4. RECEDING HORIZON CONTROL WITH NEURAL NETWORKS

Controlling nonlinear plants is in general a formidable task especially when analytical methods are used to perform the synthesis of the control action. This inherent difficulty is actually one of the main motivations advising the adoption of neural–like techniques as those investigated previously that allow to avoid cumbersome nonlinear calculations. However, the delicate and fundamental issue of the stability of the closed-loop controlled system may be better investigated in an analytical setting rather than in a heuristic one. Among various analytical approaches made available in the literature, these sections deal with the analysis of the receding horizon control strategy which was chosen for two main reasons. On the one side it is based on a well–established nonlinear programming technique that extends the classic optimal linear–quadratic regulation to nonlinear systems. On the other side, neural networks will be shown to be an invaluable tool for making receding–horizon control applicable for practical purposes. As a matter of fact, the nonlinear programming algorithm that allows to compute the receding–horizon control is so cumbersome to be incompatible with nearly all real–time applications. Only approximations via properly trained neural networks make the receding horizon technique an appealing control strategy applicable to real plants. In the following sections the basics of receding–regulators are reviewed along with their neural approximations. Numerical applications to hysteretic systems of Bouc–Wen type are eventually presented to validate the theoretical framework. As to the literature on this topic, [21] is a good (technical) reference for gaining insight into the method, [9, 10] investigate respectively on the stability and the robustness of receding-horizon controllers and [19] proposes an appealing approximate closed–form strategy to compute the receding–horizon gains. All these are referred to hereafter along with the contribution [27] which is one of the first papers introducing the idea of a combined usage of receding–horizon control and neural networks. Neural approximations of receding–horizon controllers for hysteretic systems are taken from [43, 44].

6.4.1. Elements of receding horizon control

We will follow mostly reference [27] and adhere to the discrete–time setting used therein. Original of this contribution will be the hysteretic nonlinear system to which the receding horizon control is applied. Let x_t be the state of the system at time t and consider a finite–horizon (FH) N–stage optimal control problem whose solution is the n-tuple $u_t^{FH^\circ}, \ldots u_{t+N-1}^{FH^\circ}$. Then, by definition, the first control becomes the actual control action of the receding horizon strategy, i.e. $u_t^{RH^\circ} = u_t^{FH^\circ}$. The control law that one obtains by iterating repeatedly this technique is a feedback one since the control vector $u_t^{FH^\circ}$ depends on the current state x_t. Let the system be governed by the nonlinear equation

$$x_{t+1} = f(x_t, u_t), \ t = 0, 1, \ldots, \tag{6.45}$$

where $x_t \in X \subset R^n$ and $u_t \in U \subset R^m$ are the state and control vectors respectively. It is useful to introduce first an infinite–horizon (IH) cost function that reads

$$J_{IH}(x_t, u_{t\infty}) = \sum_{i=t}^{+\infty} h(x_i, u_i), \ t \geq 0, \tag{6.46}$$

where, by definition, $u_{t\tau} = \{u_t, \ldots, u_\tau\}$. The assumptions of C^1 regularity for f and h are then classical and useful to investigate on the stabilizing properties of the controller to be designed. We want to find for every t the IH optimal feedback control law $u_t^{IH^\circ}(x_t)$ that minimizes the cost 6.46 for any x_t. The solution of the problem is computable only in the well-known case of quadratic cost and linear system and amounts to a Riccati matrix equation. To handle the nonlinear–plant non–quadratic cost case, a finite horizon (FH) problem is set forth that calls for the introduction of the FH cost

$$J_{FH}[x_t, u_{t,t+N+1}, N, h_f] = \sum_{i=t}^{t+N-1} h(x_i, u_i) + h_f(x_{t+N}), \ t \geq 0, \tag{6.47}$$

where h_f is a C^1 function that imposes a terminal penalty. The presence of the terminal penalty in the cost function is one of the two ways to impose the target of the RH control. The other one will be exploited next and consists in imposing the desired final state of the system as a further constraint to the optimization procedure. The FH problem that approximates the IH one is to find the RH control law $u_t^{RH^\circ}$ (equal to the first vector of the control sequence $u_{t(t+N-1)}^{FH^\circ}$) that minimizes the cost in equation 6.47 for the state x_t. There exist two different ways to approach the solution of the problem, and namely

- on–line computation. The problem is in fact an open–loop nonlinear programming one. Therefore, a few powerful techniques of nonlinear programming are at disposal for computing the optimal control law;
- off–line computation. A huge amount of computer memory may be necessary in this case even though the absence of on–line computing time represents a keynote of this choice.

For the existence of a receding horizon controller the following hypotheses are in order.

1. The linear system $x_{t+1} = Ax_t + Bu_t$ that linearizes the system around the origin is stabilazable;

2. the cost function $h(x, u)$ depends actually on both x and u and is coercive in the sense that there exists an increasing real positive function $r(\cdot)$ such that $h(x, u) \geq r(||x, u||), \forall x, u$;

3. the terminal cost h_f depends quadratically on the state, i.e. $h_f \in \mathcal{H}(a, P) \equiv \{h_f : h_f(x) = ax^T Px\}$, for some scalar $a > 0$ and positive definite matrix P;

4. a compact subset X_o of the state space exists such that, for every neighborhood $N \subset X_o$ of the origin, there exist a sequence of control actions $\{u_i, i = t, \ldots, t+M-1\}$ and a corresponding control trajectory $\{x_i, i = t, \ldots, t+M\}$ ending in N for every initial state $x_t \in X_o$;

5. the optimal FH feedback control $u_{FH}^o(x, i)$ is continuous with respect to the state variable x for every i.

Before stating the main existence theorem [27], the costs associated with the introduced control policies are defined as

$$
\begin{array}{ll}
J_{IH}^o(x_t) = \sum_{i=t}^{+\infty} h(x_i^{IH^o}, u_i^{IH^o}) & \text{IH optimal cost} \\
J_{RH}^o(x_t, N, h_f) = \sum_{i=t}^{+\infty} h(x_i^{RH^o}, u_i^{RH^o}) & \text{RH optimal cost} \quad (6.48) \\
J_{FH}^o(x_t, N, h_f) = \sum_{i=t}^{t+N-1} h(x_i^{FH^o}, u_i^{FH^o}) + h_f(x_{t+N}^{FH^o}) & \text{FH optimal cost}
\end{array}
$$

The following proposition holds true under hypotheses 1–5 and is fundamental.

PROPOSITION 6.4.1. *If assumptions 1–5 are satisfied, there exist a finite control horizon $\tilde{N} \geq M$, a positive scalar $\tilde{a}$ and a positive-definite symmetric matrix $P \in \mathcal{R}^{n \times n}$ such that, for every admissible terminal cost $h_f \in \mathcal{H}(a, P)$, $a > \tilde{a}$ the following properties hold.*

a) *The RH control stabilizes asymptotically the origin;*

b) *for any $N \geq \tilde{N}+1$, one has*

$$J_{RH}^o(x_t, N, h_f) \leq J_{FH}^o(x_t, N, h_f), \ \forall x_t \text{ in a proper neighborhood } \mathcal{W} \text{ of the origin;}$$

$$(6.49)$$

c) *$\forall \delta \in \mathcal{R}^+$, there exists an $N > \tilde{N}+1$ such that*

$$J_{RH}^o(x_t, N, h_f) \leq J_{IH}^o(x_t) + \delta, \ \forall x_t \in \mathcal{W}.$$

$$(6.50)$$

Thanks to Proposition 6.4.1 the existence of an horizon $\tilde{N}$ over which the RH controller stabilizes the system is guaranteed, although no practical way (rather than by trial and error) to determine the horizon is available. In view of approximating the RH control law with a backpropagation neural network it is crucial to have some robustness results ensuring that stabilizability is not lost when using a suboptimal control $\hat{u}_i^{RH}$ instead of the optimal $u_i^{RH\circ}$. The trajectory of the state under the suboptimal control is denoted by $\hat{x}_i^{RH}$. We have then the

PROPOSITION 6.4.2. *Under usual assumptions 1–5, there exist a finite $\tilde{N}$, a positive scalar $\tilde{a}$ and a positive definite symmetric matrix $P \in R^{n\times n}$ such that, for any terminal cost $h_f \in \mathcal{H}(a, P)$, $a \geq \tilde{a}$ and for any $N \geq \tilde{N}$ one has:*

a) *there exist scalars $\bar{\delta}_i > 0$ such that if $\|u_i^{RH\circ} - \hat{u}_i^{RH}\| \leq \bar{\delta}_i$, $i \geq t$, then*

$$\hat{x}_i^{RH} \in W \; \forall i > t, \; \forall x_t \in W. \tag{6.51}$$

b) *For any compact set W_d, there exist an integer $T \geq t$ and positive scalars δ_i such that if $\|u_i^{RH\circ} - \hat{u}_i^{RH}\| \leq \delta_i$, $i \geq t$, then*

$$\hat{x}_i^{RH} \in W_d \; \forall i \geq T, \; \forall x_t \in W. \tag{6.52}$$

Proposition 6.4.2 ensures the capability of the RH regulator to drive the system inside any neighborhood of the origin, provided that control errors are bounded. If the objective is to make the system asymptotically stable at the origin, a proper RH regulator may be coupled to an LQ regulator [21] that takes over when a suitable neighborhood of the origin has been reached where the linearized system is sufficiently closed to the actual one.

6.4.2. Nonlinear programming for RH control

Motivated by the stability and robustness properties of the above section, we are now interested in computing a RH controller for hysteretic systems as those introduced above. For the sake of simplicity, a quadratic cost function will be used that allows to cast the problem as a constrained quadratic programming problem as shown next. After choosing an horizon T over which the controller should operate, the continuous–time problem may be formulated as

PROBLEM 6.4.1 (Continuous RH control with terminal constraint).

$$\min_u J[x(t), t, u] = \min_u \frac{1}{2} \int_t^{t+T} [x^T(\tau)Qx(\tau) + u^T(\tau)Ru(\tau)]\, d\tau,$$
$$s.t. \qquad \dot{x} = f(x, t) + Gu, \tag{6.53}$$
$$and \qquad x(t + T) = 0,$$

where the state equation is the one described in sections 6.2–6.2.7 and no disturbance is considered when computing the control action as usual in optimal control theory.

The basic idea to solve Problem 6.4.1 is as follows. One starts by introducing a time–stepping algorithm, e.g. the Runge–Kutta method used previously or a simpler trapezoidal rule, to switch to a discrete–time, finite–dimensional problem. In particular, making repeated use of the chosen discretization scheme, the state vector at the k–th sample time of the interval $[t, t+T]$ (T being the finite horizon) may be expressed in terms of the current state $x(t)$ and the control sequence $\{u(t), \ldots, u(t+T)\}$ as

$$x(t+kh) \approx x(t) + h\left\{ \left[\sum_{i=0}^{k-1}(I+hF)^i\right]f + \sum_{i=0}^{k-1}(I+hF)^iGu[t+(k-1-i)h] \right\}, \quad 1 \le k \le N, \quad (6.54)$$

where f, G and $F \equiv \frac{\partial f}{\partial x}$ are all evaluated at the initial state $x(t)$. By introducing then the function quadratic functional $L(\tau) = x^T(t+\tau)Qx(t+\tau) + u^T(t+\tau)Ru(t+\tau)$ and using for simplicity sake the trapezoidal rule [19], the finite–horizon cost may be approximated as

$$J \approx \frac{T}{2N}[0.5L(0) + L(h) + \ldots + L((N-1)h) + 0.5L(Nh)]. \quad (6.55)$$

By plugging (6.54) in (6.55), i.e. by eliminating the dependence on $x(t+kh)$ $\forall h \in [1, N]$, one may express the cost function in terms of the current (known) state $x(t)$ and the (unknown) control n–tuple $\nu = \{u(t), u(t+h), \ldots, u(t+(N-1)h)\}$ as

$$J = \frac{1}{2}\nu^T H(x)\nu + g^T(x)\nu + q(x), \quad (6.56)$$

to be minimized with respect to ν under the final constraint that may be given the linear algebraic form

$$A^T(x)\nu = b(x). \quad (6.57)$$

The problem of minimizing the cost in (6.55) under the linear constraint 6.57 turns out to be a quadratic programming problem with equality constraint that may be solved in a finite number of steps [12].

6.4.3. Approximation by a neural network

A point made clear by the above section is that to apply the RH strategy one has to solve a nonlinear constrained optimization problem at each sample time. When the cost function is quadratic some simplifications are possible but still the method is not feasible for on–line applications because of the prohibitive computing time. Therefore, an approximate RH control law is to be derived by means of a backpropagation network trained off–line. The capability of the network to approximate the RH control law should be checked with respect to the length of the horizon as well as to output made available by proper sensors. In fact, we have tacitly assumed the state vector to be available when computing the *exact* RH control. It is however likely that this is not the case in practice. The neural network should then be trained to predict the correct control action having at disposal just a few components of the state vector. As to the Bouc–Wen model, a typical quantity one cannot rely on is the hysteretic component of the restoring force that is part of the state vector but should be considered an internal (unmeasurable) variable of the model as in classical plasticity. A simple hysteretic SDOF system is investigated next and simulations are reported concerning the RH control and its NN approximation.

6.4.4. Numerical studies

A SDOF system is investigated. Its properties were given in Tables 6.1 and 6.2. A neural network was trained to approximate the transfer function of a receding horizon controller designed according to the guidelines given in Section 6.4.2. For the time being, the state was considered available to both the RH regulator and the neural approximator and more detailed analyses are left to [43, 44]. The displacement response of Figure 6.25 as well as the hysteresis cycles of Figures 6.26 and 6.27 clearly show the potential of the method that need however to be further investigated. More insight is needed about the RH regulator as well as about its robustness properties with respect to the error coming inside due to the neural approximation.

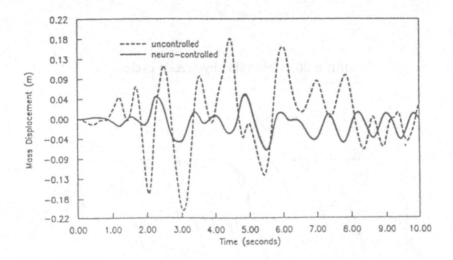

FIGURE 6.25. Uncontrolled and RH controlled displacement response

6.4.5. Concluding remarks

We have proposed the usage of a receding–horizon controller implemented on–line by means of a backpropagation neural network. The capability of handling nonlinear systems [21] along with the stabilizability and robustness properties [9, 10] make this technique an appealing alternative to more studied methodologies (especially) in the civil engineering community.

6.5. FUZZY CONTROL OF A CONTAINER-SHIP

Objective of this section is to give a rather concise exposition of fuzzy control theory presenting at the same time applications to the control of the manouvres of a

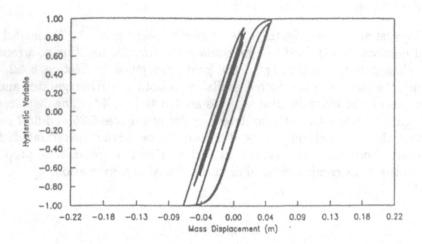

FIGURE 6.26. Controlled hysteresis cycle

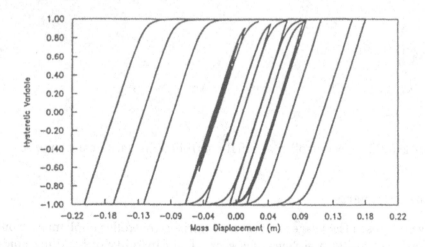

FIGURE 6.27. Uncontrolled hysteresis cycle

fast container-ship. This system has very peculiar features, e.g. a rather slow dynamics when compared to classical mechanical and civil engineering systems and, most of all, it may exhibit a chaotic behavior in the presence of high sea. For completeness sake, the analysis of the chaotic motion will be presented even though only in the uncontrolled case.

6.5.1. Basics of fuzzy control

The process of designing a fuzzy controller is somewhat similar to the human deductive process in that conclusions are successively inferred from the present knowledge. As shown in Figure 6.28, the bricks that contribute to form a fuzzy–control

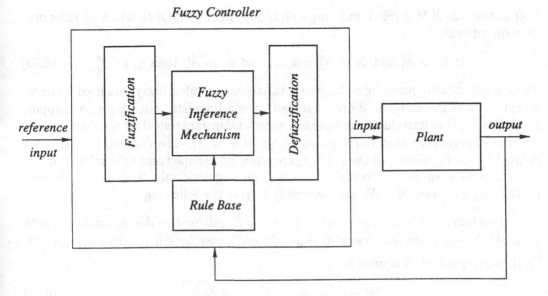

FIGURE 6.28. Fuzzy control system

environment are roughly the following ones:

1. a *rule base* that translates into the language of fuzzy sets the expert's knowledge on how to reach the goal of a performant control;
2. an *inference mechanism* which replicates the expert's job on how to interpret and apply the knowledge leading to a good control;
3. a *fuzzification interface* that decodes the inputs (measures) making them available for the inference mechanism to be used within the rules;
4. a *defuzzification interface* that converts the fuzzy output of the inference mechanism into an actual command to be applied to the plant under control.

Each of these components is next described in some detail following [28].

6.5.1.1. *The rule base*

Let a fuzzy system be a time–invariant finite–dimensional mapping between the inputs $\{u_i \in \mathcal{U}_i, \; i = 1, \ldots, n\}$ and the outputs $\{y_i \in \mathcal{Y}_i, \; i = 1, \ldots, n\}$. The domains $\mathcal{U}_i$ and $\mathcal{Y}_i$ are referred to as *universes of discourse*. Since the governing rules provided by the expert will be of linguistic type, linguistic versions of the inputs and the outputs are needed and denoted by $\tilde{u}_i$ and $\tilde{y}_i$, respectively. For instance one may have $\tilde{u}_i =$ "velocity error" or $\tilde{y}_i =$ voltage. Linguistic variables are valued in appropriate linguistic sets, e.g. let $\tilde{A}_i^j$ denote the j-th linguistic value of the linguistic variable $\tilde{u}_i$ whose universe of discourse is $\mathcal{U}_i$. Likewise the j-th linguistic value $\tilde{B}_i^j$ is defined. Typical values taken on by linguistic variables are "positive large", "zero", "negative big" and so on. The fuzzy system is then defined by means of logic implications of type

$$\text{If } (antecedent) \text{ Then } (consequent), \tag{6.58}$$

that in the case of Multiple Input Single Output system (MISO) to which all rules may be reduced reads

$$\text{If } \tilde{u}_1 \text{ is } \tilde{A}_1^j \text{ and } \tilde{u}_2 \text{ is } A_2^k \text{ and } \dots \text{ and } \tilde{u}_n \text{ is } \tilde{A}_n^l \text{ Then } \tilde{y}_q \text{ is } \tilde{B}_q^p. \tag{6.59}$$

To give quantitative meaning to linguistic variables and rules, fuzzy sets need be introduced. If $\mathcal{U}_i$ is the universe of discourse and $\tilde{A}_i^j$ is a linguistic value for $\tilde{u}_i$, a mapping $\mu : \mathcal{U}_i \to [0,1]$ is introduced and named *membership function* that accounts for the "degree–of certainty" that the linguistic value of $\tilde{u}_i$ is $\tilde{A}_i^j$. Membership functions are defined by the experience of the analyst, are often bell–shaped and take values close to one and zero when we are "certain" that a given linguistic value is or is not attained by the linguistic variable. We are now ready to give the following

DEFINITION 6.5.1. *(Fuzzy set). Given a linguistic variable $\tilde{u}_i$ with linguistic value $\tilde{A}_i^j$ defined on the universe of discourse $\mathcal{U}_i$ and a membership function $\mu_{A_i^j}(u_i) : \mathcal{U}_i \to [0,1]$, a fuzzy set A_i^j is defined as*

$$A_i^j = \left\{ \left(u_i, \mu_{A_i^j}(u_i) \right), \ u_i \in \mathcal{U}_i \right\}, \tag{6.60}$$

i.e. the set of pairs of variables and associated membership functions for all possible values taken on in the universe of discourse.

In order to perform fuzzy calculations, the algebra of fuzzy sets must be introduced and notions as subset, union and intersection given.

DEFINITION 6.5.2. *(subset). With the above definitions and notations, the fuzzy set A_i^1 is said to be a fuzzy subset of A_i^2 if $\mu_{A_i^1} \leq \mu_{A_i^2}$, $\forall u_i \in \mathcal{U}_i$. In symbols, we write $A_i^1 \subset A_i^2$.*

DEFINITION 6.5.3. *(intersection). The intersection of fuzzy sets A_i^1 and A_i^2 is the fuzzy set $A_i^1 \cap A_i^2$ whose membership function is either*

$$\begin{aligned} Minimum & \quad \mu_{A_i^1 \cap A_i^2} = \min\{\mu_{A_i^1}, \mu_{A_i^2}\}, \ or \\ Algebraic\ product & \quad \mu_{A_i^1 \cap A_i^2} = \{\mu_{A_i^1} \mu_{A_i^2}\}. \end{aligned} \tag{6.61}$$

Any of the two choices above are usually denoted by a star, i.e. $x * y = \min\{x, y\}$ or $x * y = xy$. As in classical set theory, the intersection is used to denote the and operation.

DEFINITION 6.5.4. *(union) The union of fuzzy sets A_i^1 and A_i^2 is the fuzzy set $A_i^1 \cup A_i^2$ whose membership function is either*

$$
\begin{aligned}
Maximum \quad & \mu_{A_i^1 \cup A_i^2} = \max\{\mu_{A_i^1}, \mu_{A_i^2}\}, or \\
Algebraic\ sum \quad & \mu_{A_i^1 \cup A_i^2} = \{\mu_{A_i^1} + \mu_{A_i^2} - \mu_{A_i^1}\mu_{A_i^2}\}.
\end{aligned}
\tag{6.62}
$$

Any of the two choices above are usually denoted by $\oplus$, i.e. $x \oplus y = \max\{x, y\}$ or $x \oplus y = x + y - xy$. As in classical set theory, the intersection is used to denote the or operation.

The operations of union and intersection may be computed between two fuzzy sets that share the same universe of discourse. An operation that is made between fuzzy sets not necessarily defined on the same universe of discourse is the Cartesian product that is a fuzzy set whose membership function is the iterated $*$ product of the membership functions of the single fuzzy sets.

6.5.1.2. *Fuzzification*

Fuzzification is about how the numeric inputs $u_i \in \mathcal{U}_i$ are converted into fuzzy sets to be used by the inference mechanism. The set of all fuzzy sets that make sense on $\mathcal{U}_i$ is denoted by $\mathcal{U}_i^*$. The fuzzification operator $\mathcal{F}$ is a mapping $\mathcal{F} : \mathcal{U}_i \to \mathcal{U}_i^*$ that may be formally written as $\mathcal{F}(u_i) = \hat{A}_i^{fuz}$. The most widely used scheme is the singleton fuzzification which produces a fuzzy set $\hat{A}_i^{fuz} \in \mathcal{U}_i^*$ whose membership function is given as

$$
\mu_{\hat{A}_i^{fuz}} = \begin{cases} 1 & x = u_i \\ 0 & else \end{cases}
\tag{6.63}
$$

Using the singleton fuzzification amounts to neglecting the effect of noise which disturbs measurements. If noise is to be considered, a Gaussian fuzzification may be used using bell shaped membership functions.

6.5.1.3. *The inference mechanism*

The core of the fuzzy control method is represented by the inference mechanism that serve mainly two purposes, i.e.

1. based on the inputs u_i, $i = 1, \ldots, n$, determine the importance of each rule in the current situation. This operation is referred to as *matching*.
2. The true *inference step* that consists in drawing conclusions from the available input.

6.5.1.4. *Matching*

Let us start by defining the fuzzy set of the antecedent of the i–th rule, i.e. $A_1^j \times A_2^k \times \cdots A_n^l$, and that of the fuzzified inputs, i,e, $\hat{A}_1^{fuz} \times \hat{A}_2^{fuz} \times \cdots \hat{A}_n^{fuz}$. Combining fuzzy sets from fuzzification and those from the antecedent of the rule fuzzy sets

$\hat{A}_1^j, \hat{A}_2^k, \ldots, \hat{A}_n^l$ with membership function

$$\mu_{\hat{A}_i^k}(u_i) = \mu_{A_i^k} * \mu_{\hat{A}_i^{fuz}}(u_i), \ i = 1, \ldots, n. \tag{6.64}$$

Then the membership function of the Cartesian product $\hat{A}_1^j \times \hat{A}_2^k \times \cdots \times \hat{A}_n^l$ is formed as

$$\overline{\mu}_i(u_1, u_2, \ldots, u_n) = \mu_{\hat{A}_1^j}(u_1) * \mu_{\hat{A}_2^k}(u_2) * \cdots \mu_{\hat{A}_n^l}(u_n), \tag{6.65}$$

that, when using singleton fuzzification, yields

$$\overline{\mu}_i(\overline{u}_1, \overline{u}_2 \ldots, \overline{u}_n) = \mu_{\hat{A}_1^j}(\overline{u}_1) * \mu_{\hat{A}_2^k}(\overline{u}_2) * \cdots \mu_{\hat{A}_n^l}(\overline{u}_n), \ for \ \overline{u}_i = u_i, \tag{6.66}$$

and $\overline{\mu}_i(\overline{u}_1, \overline{u}_2 \ldots, \overline{u}_n) = 0$ for $\overline{u}_i \neq u_i$. The matching is then ended by using

$$\mu_i(u_1, u_2, \ldots, u_n) = \mu_{A_1^j}(u_1) * \mu_{A_2^k}(u_2) * \cdots \mu_{A_n^l}(u_n) \tag{6.67}$$

to represent the certainty that the antecedent rule i matches the input information.

6.5.1.5. *Inference*

For each rule, say the i-th, the implied fuzzy set $\hat{B}_q^i$ is determined. $\hat{B}_q^i$ specifies to which extent we are certain that the output should take a specific value (*crisp*) $y_q \in \mathcal{Y}_q$, taking into account the i-th rule only. The relevant membership function reads

$$\mu_{\hat{B}_q^i}(y_q) = \mu_i(u_1, u_2, \ldots, u_n) * \mu_{B_q^p}. \tag{6.68}$$

The overall implied fuzzy set $\hat{B}_q$ is the one whose membership function is given as

$$\mu_{\hat{B}_q}(y_q) = \mu_{\hat{B}_q^1}(y_q) \oplus \mu_{\hat{B}_q^2}(y_q) \oplus \cdots \mu_{\hat{B}_q^R}(y_q), \tag{6.69}$$

that is the conclusion based on all rules in the rule base. The inspiring principle of (6.69), that is often referred to as sup–star inference rule, is that *we can be no more certain about conclusions than we are about premises.*

6.5.1.6. *Defuzzification*

Defuzzification is the process of converting fuzzy conclusions coming from the inference mechanism into numerical values for controlling the system. A number of approaches are available for this purpose, e.g. the max criteria, the mean of maximum, the center of area, the center average and the center of gravity (COG) that is reported hereafter since it has been used in the numerical simulations.

6.5.1.7. *Center of gravity defuzzification*

A crisp output y_q^{crisp} is determined according to

$$y_q^{crisp} = \frac{\sum_{i=1}^{R} c_q^i \int_{\mathcal{Y}_q} \mu_{\hat{B}_q^i}(y_q) dy_q}{\sum_{i=1}^{R} \int_{\mathcal{Y}_q} \mu_{\hat{B}_q^i}(y_q) dy_q}, \tag{6.70}$$

where c_q^i is the center of area of the membership function of B_q^p associated with the implied fuzzy set $\hat{B}_q^i$.

6.5.2. Multivariable fuzzy control of a container-ship

Multivariable fuzzy control is hereafter applied to a nonlinear model of a container-ship. Before presenting a few numerical results on the performance of such control strategy a detailed analysis is given showing that the system is prone to exhibit chaotic effects that might tremendously affect the performance of the controller.

6.5.2.1. *Introductory remarks*

Previous studies evidenced that the dynamic behavior of several ships character-ized by a low metacentric height cannot be adequately described in terms of the simpli-fied models of linear and decoupled type, which are generally used in the hydrodynamic and naval architecture literature [37, 5]. It has in fact been clearly demonstrated that for such vessels, including container-ships, ro–ro ships, high speed ferries, naval vessels and fishing boats, non linear coupled surge–yaw–sway–roll can supply a better knowl-edge about the precise interaction between roll and other motions. It has also been recently shown that non linear rolling of a fishing boat in irregular waves, analyzed by a single–degree–of–freedom model, can induce quite extreme effects including ship capsizing. Even if these extreme effects, including evident chaotic phenomena, have been often observed in the practice of ship operations at sea, an inadequate theoret-ical investigation has prevented to find out, in the past years, plausible explanations for such phenomena. In fact, the availability of methods and of numerical algorithms capable to describe and efficiently characterize chaos in dynamical systems, has taken place only in the last decade.

The description of the interaction between ship motions and sea waves has been tra-ditionally carried out on the basis of a linear superposition principle, from which it is mathematically impossible to deduce chaotic effects.

The present contribution is aimed to show further evidence of such effects, by de-veloping an analysis of a four degree–of–freedom container-ship model, perturbed by stochastic sea waves. The rough sea–ship interactions leading to possible chaotic ef-fects in the roll as well as in the other motions are investigated by using a numerical method, which allows to determine Lyapunov exponents from time series. Also, time domain simulations and phase space representations are used to gain insight into the behavior of the system.

6.5.2.2. *Modeling of ship motions in waves*

The mathematical model of the container-ship considered herein is described in detail in [38]. It is herewith considered a stochastic extension of such model capable to describe the ship response in irregular sea waves, which is expressed by the following non linear equations

$$
\begin{aligned}
m(\dot{u} - vr - x_G r^2 + z_G pr) &= X + X_w \\
m(\dot{v} + ur - z_G \dot{p} + x_G \dot{r}) &= Y + Y_w \\
I_{zz}\dot{r} + mx_G(ur + \dot{v}) &= N + N_w \\
I_{xx}\dot{p} - mz_G(ur + \dot{v}) &= K + K_w - \rho g D G_z(\varphi)
\end{aligned}
\tag{6.71}
$$

where D is the displacement, g the gravity constant, ρ the water mass density, $G_z(\varphi)$ is the action of the rightening arm that depends on the roll angle φ, while $(x_G, 0, z_G)$ are the coordinates of the mass center. The mass is denoted by m whereas I_{xx} and I_{zz} are the inertial moments about x and z, respectively. The linear velocity of surge and sway are u and v and the angular ones of yaw and roll are r e p. According to [29], the rightening arm action may be expressed as:

$$
G_z(\varphi) = \sin\varphi \left(GM + \frac{BM}{2} \tan^2 \varphi \right),
\tag{6.72}
$$

where GM is the metacentric height and BM is the distance from the center of buoyancy to the metacenter.

The terms X, Y, and N, K denote the forces along the x and y axes and the moments around the z and y axes in the ship fixed system, respectively. These quantities take into account the hydrodynamic effects from hull movements, forces exerted on the ship by the rudder and by the propulsion system, and the effects induced by sea waves, wind and currents.

Forces and moments X, Y, N and K that appear in (6.71) are usually polynomials whose independent variables are the states of the system and the rudder angle δ [18]. Their specific expressions for the single-screw high-speed container–ship used herein, are given in [5] whereas 6.5 presents a list of its principal data. A slightly simplified representation of the hydrodynamic forces and moments for the container-ship described

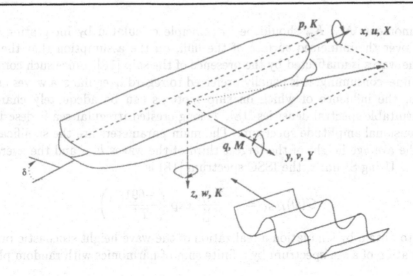

FIGURE 6.29. Schematic of the ship and notations

in [5] is used in [37]:

$$X = R(u) + (1 - t_o)T + X_{\dot{u}}\dot{u} + X_v v + X_{vv}v^2$$
$$+ X_{vr}vr + X_{v\varphi}v\varphi + X_\varphi\varphi + X_{\varphi\varphi}\varphi^2 + X_\delta\delta$$
$$+ X_{\delta\delta}\delta^2,$$

$$Y = Y_v v + + Y_{vv}v^2 + Y_{|v|v}\,|\,v\,|\,v + Y_p p + Y_\varphi\varphi$$
$$+ Y_{vvv}v^3 + Y_{rrr}r^3 + Y_{vvr}v^2 r + Y_{vrr}vr^2$$
$$+ Y_{vv\varphi}v^2\varphi + Y_{v\varphi\varphi}v\varphi^2 + Y_{rr\varphi}r^2\varphi$$
$$+ Y_{r\varphi\varphi}r\varphi^2 + Y_\delta\delta + Y_{\delta\delta}\delta^2,$$

$$N = N_v v + N_r r + N_p p + N_\varphi\varphi + N_{vvv}v^3$$
$$+ N_{rrr}r^3 + N_{vvr}v^2 r + N_{vrr}vr^2 + N_{vv\varphi}v^2\varphi$$
$$+ N_{v\varphi\varphi}v\varphi^2 + N_{rr\varphi}r^2\varphi + N_{r\varphi\varphi}r\varphi^2$$
$$+ N_\delta\delta + N_{\delta\delta}\delta^2,$$

$$(6.73)$$

$$K = K_v v + K_r r + K_p p + K_\varphi\varphi + K_{vvv}v^3$$
$$+ K_{rrr}r^3 + K_{vvr}v^2 r + K_{vrr}vr^2 + K_{vv\varphi}v^2\varphi$$
$$+ K_{v\varphi\varphi}v\varphi^2 + K_{rr\varphi}r^2\varphi + K_{r\varphi\varphi}r\varphi^2$$
$$+ K_\delta\delta + K_{\delta\delta}\delta^2.$$

Here, δ is rudder angle, T propeller thrust, t_o the thrust deduction factor, and $R(u)$ the hull resistance. The coefficients of these equations are constants, referred to as hydrodynamic derivatives, and represented by terms like $X_{|u|u} = \frac{\partial^2 X}{\partial|u|\partial u}$, $Y_v = \frac{\partial Y}{\partial v}$, $N_{vrr} = \frac{\partial^3 N}{\partial v\partial r^2}$ The stochastic components of forces and moments terms are denoted by a subscript w. In fact, the evaluation of sea waves induced perturbing forces X_w,

Y_w and moments N_w, K_w should be in principle calculated by integrating the wave pressure over the immersed surface of the hull, on the assumption that the pressure within the waves is unaffected by the presence of the ship [18]. Since such computation is quite time–consuming, it is usually preferred to regard irregular sea waves as random processes, the influence of which on ship motions can be adequately characterized through suitable spectral densities [18]. A long crested irregular sea is described by a one–dimensional amplitude spectrum. The main parameters are the significant wave–height, the average height of the largest third of the waves $h_{1/3}$ and the average wave period T_w. Using SI units, the ISSC spectrum [18] is

$$G(\omega)_{ISSC} = \frac{173h_{1/3}^2}{\omega^5 T_w^4} \exp\left(\frac{-691}{T_w^4}\omega^4\right) \tag{6.74}$$

In order to obtain by simulation a realization of the wave height stochastic process, an approximation of a sea spectrum by a finite sum of harmonics with random phases has been used:

$$z(t) = \sum_{i=1}^{n} a_i \sin(\omega_{e,i} t + \phi_i). \tag{6.75}$$

The individual amplitudes of the sinosoids are conveniently taken as median points between frequencies where the spectrum amplitude is significantly different from zero and the phases ϕ_i are random numbers uniformly distributed between zero and 2π. The coefficients a_i are calculated from energy considerations according to [18] and $\omega_{e,i}$ are the encounter angular frequencies between ship and waves direction, which are related to ω, to ship forward speed U and to the angle of encounter χ by

$$\omega_e = \omega\left(1 - \omega\frac{U}{g}\cos\chi\right). \tag{6.76}$$

The corresponding effects in terms of wave induced forces and moments can be obtained by multiplying the wave height $z(t)$ given by 6.75 by a suitable gain coefficient, which is assumed, in a first approximation, to depend on the frequency of encounter ω_e and on the angle of encounter χ. The roll moment can be written for example as

$$K_w = \gamma_K(\omega_e, \chi)z \tag{6.77}$$

Similar expressions can be assumed for the other forces and moments X_w, Y_w and N_w.

6.5.2.3. Preliminary simulations

A number of simulations have been carried out at different sea–waves spectral intensities and different angles of encounter between ship and waves. During such simulations, only the interaction of the ship with the sea waves has been considered, i.e. the ship has been regarded as navigating at nominal forward speed U along a prescribed fixed course and subject only to sea waves induced forces and moments. This means that the components depending on the rudder angle δ are set to zero in (6.73).

In figure 6.30, an example of time series of roll, roll rate, yaw and sway is reported. Such responses refer to a situation where the ship is navigating at a nominal forward speed of 12 m/sec, while the maximum disturbance is on the roll motion. The wave height $h_{1/3}$ and the average wave period T_w are respectively assumed to be 5 m and 10 sec; an angle of encounter $\chi = 90$ degrees has also been considered, corresponding to a beam–sea condition. It can be noticed that owing to coupling, all the motion components are perturbed, even if the yaw and sway forcing excitations are relatively small with respect to the corresponding one related to roll.

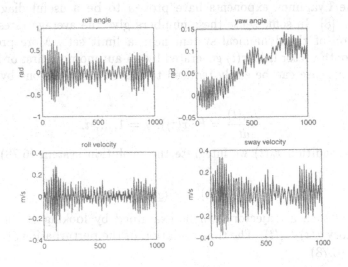

FIGURE 6.30. Plot of roll angle and velocity and sway velocity and yaw angle

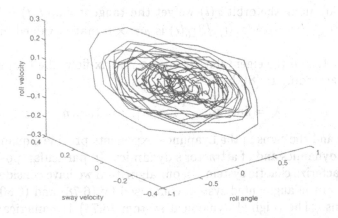

FIGURE 6.31. Plot of a typical phase section

6.5.2.4. *Analysis of chaotic effects*

A qualitative evaluation of a possibly chaotic behavior has been first carried out by drawing phase sections of the simulated trajectories. In figure 6.31, a 3D phase–space section is illustrated from which the existence of an attractor limit cycle is evidenced. In order to be able to achieve a quantitative characterization of the ship chaotic behavior, a numerical method based on Lyapunov exponent computation has been applied. One of the most used characterization of the behavior of a dynamical system of dimension n is its Lyapunov spectrum of characteristic exponents [33]. In particular, the Lyapunov exponents have proven to be a useful diagnostic tool for chaotic system [8]. In some sense these numbers give the average rates of expansions and contraction of the dynamical system near a limit set. More precisely we can consider a smooth vector field $\underline{x}(t)$ generated by an autonomous first order system (the non autonomous case can be easily reduced to the autonomous one by appending an extra state),

$$\frac{dx_i(t)}{dt} = F_i(\underline{x}(t)), \ i = 1, ..., n. \tag{6.78}$$

For an initial condition $\underline{x}_0(t)$ we integrate the evolution system (6.78) to obtain the trajectory

$$\underline{x}(t) = \varphi_t(\underline{x}_0(t)). \tag{6.79}$$

The stability of such a trajectory can be examined by looking at the behavior of a nearby trajectory $\underline{x}(t) + \underline{\xi}(t)$. The time evolution of the perturbation $\underline{\xi}(t)$ is then found by linearizing (6.78)

$$\frac{d\underline{\xi}(t)(t)}{dt} = J(\underline{x}(t)) \cdot \underline{\xi}(t), \tag{6.80}$$

where $J(\underline{x}(t)) = \partial F / \partial \underline{x}(t)$ is the Jacobian matrix of F.

Integrating (6.80) along the orbit $\underline{x}(t)$ we get the tangent map $\underline{\xi}(t) = T_{\underline{x}_0(t)}(t)\underline{\xi}_0(t)$ in which the flow $T_{\underline{x}_0(t)}(t) = \partial\phi_t(\underline{x}_0(t))/\partial\underline{x}_0(t)$ is a $n \times n$ matrix valued matrix.

Let $\{\sigma_i(t)\}$, $i = 1, ..., n$ the eigenvalues of such a matrix flow, the Lyapunov exponents λ_i are defined, according to [8], by

$$\lambda_i = \text{limsup}_{t \to \infty} \frac{1}{t} \log |\sigma_i|, \ i = 1, ..., n. \tag{6.81}$$

The magnitude and the signs of the Lyapunov exponents provide a quantitative picture of the system's dynamics and of attractor's dynamics. In particular, positive Lyapunov exponents characterize chaotic system. In our algorithm we have considered the numerical integration of the augmented system composed by (6.78) and (6.80) with suitable initial conditions. The original dynamical system (6.71) is numerically non stiff so that classical Runke-Kutta schemes can be adopted, while for the tangential flow the implicit Euler method has been used. In order to find the largest Lyapunov exponent, say λ_1, a time interval τ and an initial perturbation $\underline{\xi}_0(t)$ (normalized to 1), and the solution of (6.80) has been computed for $t = k\tau$, where k is a count parameter. Every τ_1 seconds we normalize the solution to prevent overflow situations. The normalizing factors provide us an estimate of the number λ_1 [8].

Let $\underline{\xi}^1(t)$ be the trajectory with the normalized restart, then

$$\lambda_1 \sim \frac{1}{k\tau} \sum_i \log \| \underline{\xi}^1(i\tau_1) \|, \tag{6.82}$$

for sufficiently large k. We note that the procedure starts after an initial time delay, in order to consider the initial transient of the system.

In figure 6.32, we have shown the behavior of the discrete estimates (6.82) to the

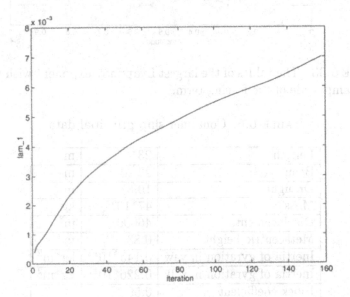

FIGURE 6.32. Time evolution of the discrete sum (6.82) representing the largest Lyapunov exponent λ_1 for a typical forcing simulation

largest Lyapunov exponent λ_1 for a typical simulation with wave sea perturbation. In this simulation the sign of the final numerical exponent is definitely positive and monotonically increasing. In figure 6.33, we have plotted the evolution of the exponent λ_1 with respect to the amplitude of the roll forcing term γ_K varying in the range $[10^{-6}, 10^{-4}]$. For this last figure the Lyapunov exponent λ_1 is computed as an average value from several simulations with the same perturbation but different initial conditions, as in a Monte Carlo simulation context. It can be noticed that the chaotic behavior tends to increase monotonically after reaching a certain threshold amplitude value. This effect seems to be in a good agreement with capsizing phenomena.

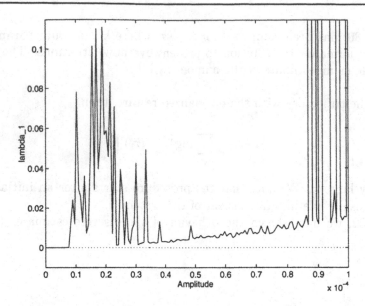

FIGURE 6.33. The values of the largest Lyapunov exponent with respect to the amplitude of the forcing term.

TABLE 6.5. Container ship principal data

Length	231	m
Beam	32	m
Draught	10.81	m
Mass	$47*10^6$	kg
Displacement	46000	m^3
Metacentric height	0.83	m
Inertia of gyration in yaw	$0.145*10^{12}$	$kg*m^2$
Inertia of gyration in roll	$0.226*10^{11}$	$kg*m^2$
Block coefficient	0.56	
Propeller diameter	8	m
Rudder area	60	m^2
Rudder rate (one pump)	2.3	deg/s
Cruising speed	12.7	m/s

6.5.2.5. *Fuzzy control of multivariable ship motion*

One of the most interesting and challenging tasks in the area of ship motion control concerns ship rudder roll damping (RRD). The philosophy behind RRD is that the rudder can be used as the only actuator to reduce roll motion and maintain the ship course at the same time. This result can be achieved in principle by compensating the disturbing effects induced by sea waves through a low frequency rudder action for the yaw and a high frequency one for the roll. The analysis of the RRD problem was made in [41] via PID regulation, in [30] by means of a frequency domain autopilot, in [15] using an LQG controller and in [38] via artificial neural networks. These contributions shows the potential of the RRD technique even though are mostly based on linearized weakly–coupled ship models. On the other hand, ships with low metacentric height cannot be adequately investigated under this simplifying assumptions [5] and need a fully–coupled nonlinear model. The objective of the forecoming sections is therefore to design a multivariable fuzzy controller operating on the coupled nonlinear system. Further details may be found in [39] and [40].

6.5.3. The fuzzy controller

Fuzzy controllers (FC) are well suited for the problem above since they do not critically depend on the ship mathematical model and can also cope quite well with nonlinear plants. The general structure of the controlled plant was already presented in figure 6.28 and the assumed linguistic values of the variables are simply Positive (P), Negative (N) and Zero (Z). Figure 6.34 presents the membership functions for the input used in the numerical simulations, i.e. ε_ψ, r, ε_φ, p that denote respectively the heading error, the yaw rate, the roll error and the roll rate. Following [36] a bell–shape function is used that reads

$$f_i(x, \alpha_i, \beta_i, \gamma_i) = \frac{1}{1 + \left(\frac{x-\gamma_i}{\alpha_i}\right)^{2\beta_i}}, \quad i = 1, \dots, 4. \tag{6.83}$$

The inference procedure is based on a suitable rule–base system which can be easily deduced in the case of SISO system whereas may lead to serious problems in the case of multivariable plants. In fact, the number of possible rules may be subjected to an exponential growth with respect to the input/output dimension. It is therefore mandatory to prune the rule tree by eliminating all possible redundancies. In our case [39, 40] Sugeno–type rules are used that extend the ones of the single–variable autopilot presented in [36]. The rules are in fact all possible combinations that derive by the pairs (ε_ψ, r), (ε_φ, p), (ε_ψ, p) and (ε_φ, r) amounting to 36 rules. Cross coupling between yaw and roll motions is therefore included herein whereas this was not the case in [36]. By denoting by δ the rudder angle, the (ε_φ, r) rules read

$$
\begin{array}{llll}
\text{1) If} & \varepsilon_\varphi \text{ is } N \text{ and } r \text{ is } N & \text{Then} & \delta_1 = a_1\varepsilon_\psi + b_1 r + c_1\varepsilon_\varphi + d_1 p + g_1 \\
\text{2) If} & \varepsilon_\varphi \text{ is } N \text{ and } r \text{ is } Z & \text{Then} & \delta_2 = a_2\varepsilon_\psi + b_2 r + c_2\varepsilon_\varphi + d_2 p + g_2 \\
\text{3) If} & \varepsilon_\varphi \text{ is } N \text{ and } r \text{ is } P & \text{Then} & \delta_3 = a_3\varepsilon_\psi + b_3 r + c_3\varepsilon_\varphi + d_3 p + g_3 \\
\text{4) If} & \varepsilon_\varphi \text{ is } Z \text{ and } r \text{ is } N & \text{Then} & \delta_4 = a_4\varepsilon_\psi + b_4 r + c_4\varepsilon_\varphi + d_4 p + g_4 \\
\text{5) If} & \varepsilon_\varphi \text{ is } Z \text{ and } r \text{ is } Z & \text{Then} & \delta_5 = a_5\varepsilon_\psi + b_5 r + c_5\varepsilon_\varphi + d_5 p + g_5 \\
\text{6) If} & \varepsilon_\varphi \text{ is } Z \text{ and } r \text{ is } P & \text{Then} & \delta_6 = a_6\varepsilon_\psi + b_6 r + c_6\varepsilon_\varphi + d_6 p + g_6 \\
\text{7) If} & \varepsilon_\varphi \text{ is } P \text{ and } r \text{ is } N & \text{Then} & \delta_7 = a_7\varepsilon_\psi + b_7 r + c_7\varepsilon_\varphi + d_7 p + g_7 \\
\text{8) If} & \varepsilon_\varphi \text{ is } P \text{ and } r \text{ is } Z & \text{Then} & \delta_8 = a_8\varepsilon_\psi + b_8 r + c_8\varepsilon_\varphi + d_8 p + g_8 \\
\text{9) If} & \varepsilon_\varphi \text{ is } P \text{ and } r \text{ is } P & \text{Then} & \delta_9 = a_9\varepsilon_\psi + b_9 r + c_9\varepsilon_\varphi + d_9 p + g_9
\end{array}
\tag{6.84}
$$

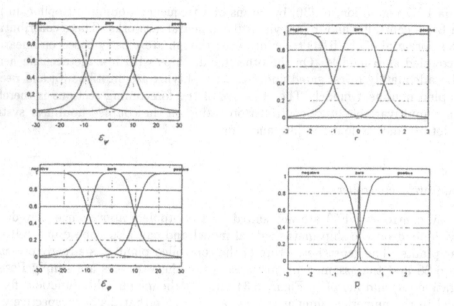

FIGURE 6.34. Fuzzy sets for FC input variables (heading error, yaw rate, roll error and roll–rate

The other three groups of rules are of course of the same type. Defuzzification of the output variable (the rudder angle) was made thanks to the simple relation

$$\delta = \frac{\sum_{i=1}^{36} w_i \delta_i}{\sum_{i=1}^{36} w_i},$$

(6.85)

where the coefficients w_i are the weighting factors of the antecedent of the i-th rule having taken into account the fuzzy set intersection operator. In order to ensure the performance of the controller over a wide range of operating conditions, it is necessary to develop a learning procedure according to which the 12 parameters in (6.83) and the 180 ones appearing in the rules (6.84). If ϑ is the vector grouping all the unknowns, a quadratic error function is set forth as

$$J(\vartheta) = \frac{1}{2} \sum_{j=1}^{N} [\hat{x}(j) - x(j)]^T [\hat{x}(j) - x(j)],$$

(6.86)

where $\{\hat{x}(j)\}_{j=1}^{N}$ is a time record of a selected training vector containing the ship variables under control as obtained by a reference controller whereas $\{x(j)\}_{j=1}^{N}$ are the corresponding values as given by the FC. The minimization was made by simulated annealing to avoid local minima. Figure 6.35 gives a schematic of the fuzzy ship control system.

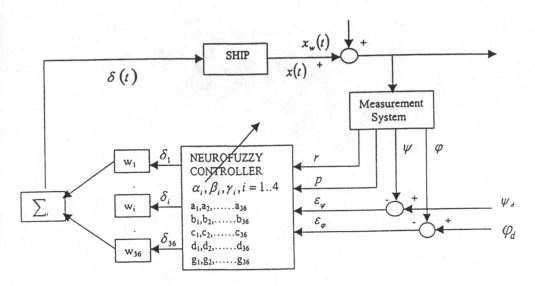

$$\text{FIGURE 6.35. Architecture of the fuzzy ship control system}$$

6.5.3.1. *Simulation results*

The same container-ship studied previously is now considered to evaluate the behavior of the fuzzy RRD controller. A wide range of sea conditions have been simulated with particular attention to the case of an angle of encounter equal to 90 degrees for which the effect of waves on the roll motion is maximum. The reference control is an LQ autopilot designed according to the linearized plant. Thanks to the time series generated using the LQ regulator, the parameters of the fuzzy rules were tuned. Owing to the extremely high number of parameters to be determined the optimization procedure was quite time consuming and sometimes stopped before reaching the global minimum. Figure 6.36 shows some selected elements of fuzzy inference system matrix. A comparison between the reference LQ controller and the fuzzy controller is made in Figures 6.37 and 6.38 where the response to a 10 degrees yaw angle demand is illustrated. The container ship was simulated at a velocity of 15 m/s with an angle of encounter of 90 degrees between ship and waves. The better dynamics of the FC autopilot may be visually appreciated. This excellent behavior is confirmed also in the course–changing simulation shown in Figures 6.39 and 6.40.

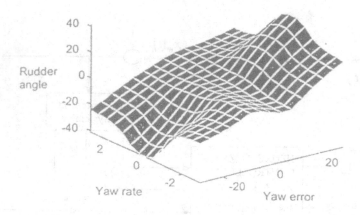

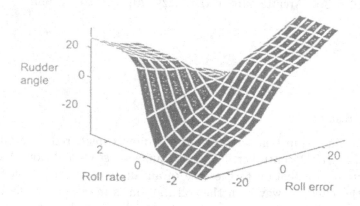

FIGURE 6.36. Output surfaces corresponding to selected elements of FIS (Fuzzy Inference System) matrix

6.6. CONCLUSIONS

This chapter has given an overview of possible applications of neural networks and fuzzy logic in the area of active control of system and structures. Regulating open-loop unstable rigid systems and flexible nonlinear ones is known to be one of the most challenging problems that one might face in the area of active control. Neural networks

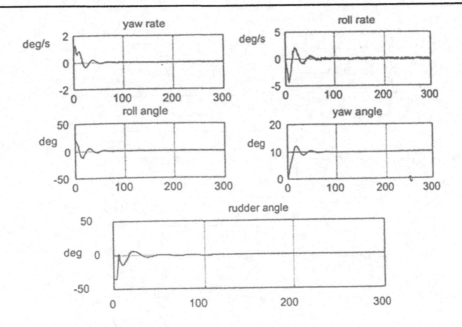

FIGURE 6.37. Course–keeping and roll–damping. LQ control with significant wave height $h = 5$ m and wave period $T = 10$ s

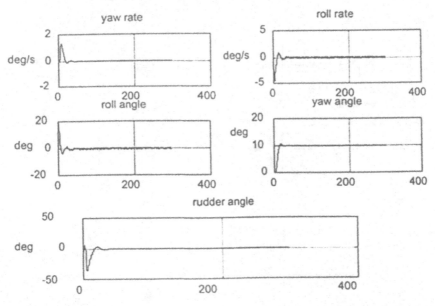

FIGURE 6.38. Course–keeping and roll–damping. Fuzzy control with significant wave height $h = 5$ m and wave period $T = 10$ s

and fuzzy logic were shown to be quite adequate for this purpose, especially in view of

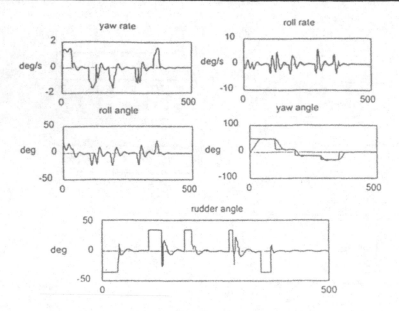

FIGURE 6.39. Course–changing and roll–damping. LQ control with significant wave height $h = 5$ m and wave period $T = 10$ s

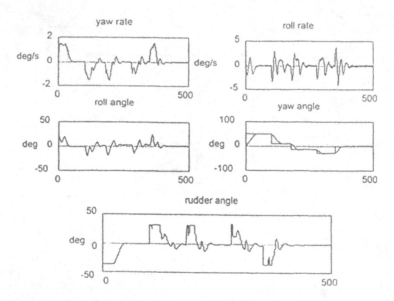

FIGURE 6.40. Course–changing and roll–damping. Fuzzy control with significant wave height $h = 5$ m and wave period $T = 10$ s

ad–hoc algorithms that were conceived and applied. Many details were given on the control algorithms as well as on the results from numerical and experimental tests.

ACKNOWLEDGMENT

The author thanks Prof. J. Ghaboussi of the University of Illinois at Urbana–Champaign and Prof. R. Scattolini of the University of Pavia for precious suggestions on some parts of the chapter. Prof. Z. Waszczyszyn of the Cracow University of Technology who made this contribution possible is gratefully acknowledged.

REFERENCES

1. B. ARMSTRONG-HÉLOUVRY, P. DUPONT, AND C. CANUDAS DE WIT, *A survey of models, analysis tools and compensation methods for the control of machines with friction*, Automatica, 30 (1994), pp. 1083–1138.
2. T. ASH, *Dynamics Node Creation in Backpropagation Networks*, ICS Rep. 8901, Univ. of California, San Diego (Ca), 1989.
3. T. BABER AND Y.-K. WEN, *Random vibrations of hysteretic degrading systems*, ASCE J. Engrg. Mech. Div., 107(6) (1981), pp. 1069–1087.
4. K. BANI-HANI AND J. GHABOUSSI, *Nonlinear structural control using neural networks*, ASCE J. Engrg. Mech. Div., 124(3) (1998), pp. 319–327.
5. M. BLANKE AND A. JENSEN, *Dynamic properties of container vessel with low metacentric height*, Trans. Inst. Meas. and Cont., 19(2) (1997), pp. 78–93.
6. E. BLUM AND L. LI, *Approximation theory and feedforward networks*, Neural Networks, 4 (1991), pp. 511–515.
7. H. CHEN, K. TSAI, G. QI, J. YANG, AND F. AMINI, *Neural network for structure control*, ASCE J. Computing in Civil Engrg., 9(2) (1995), pp. 168–176.
8. L. CHUA AND S. PARKER, *Chaos: a tutorial for engineers*, Proc. IEEE, 78(3) (1987), pp. 982–1008.
9. G. DE NICOLAO, L. MAGNI, AND R. SCATTOLINI, *On the robustness of receding-horizon control with terminal constraints*, IEEE Trans. Autom. Control, 41(3) (1996), pp. 451–453.
10. G. DE NICOLAO AND S. STRADA, *On the stability of receding-horizon lq control with zero-state terminal constraint*, IEEE Trans. Autom. Control, 42(2) (1997), pp. 257–260.
11. L. FARAVELLI AND P. VENINI, *Active structural control by neural networks*, Journal of Structural Control, 1(1-2) (1994), pp. 79–101.
12. R. FLETCHER, *Practical Methods of Optimization, 2nd edn*, John Wiley and Sons, New York, 1989.
13. J. GHABOUSSI AND A. JOGHATAIE, *Active control of structures using neural networks*, ASCE J. Engrg. Mech. Div., 121(4) (1995), pp. 555–567.
14. K. HORNIK, *Approximation capabilities of multilayered feedforward networks*, Neural Networks, 4 (1991), pp. 251–257.
15. C. KALLSTRÖM AND S. W.L., *An integrated rudder control system for roll damping and maintenance*, Proc. IX Ship Cont. Syst. Symposium, Bethesda (1990), pp. 278–296.
16. H. KWAKERNAAK AND R. SIVAN, *Linear Optimal Cintrol*, John Wiley, New York, 1972.

17. W. KWON AND A. PEARSON, *A modified quadratic cost problem and feedback stabilization of a linear system*, IEEE Trans. Automat. Contr., 22 (1997), pp. 838–842.

18. E. LEWIS, *Principles of Naval Architecture*, Soc. of Naval Arch. and Marine Engrg., New York, 1988.

19. P. LU, *Approximate nonlinear receding-horizon control laws in closed form*, Int. J. Control, 71(1) (1998), pp. 19–34.

20. MATLAB, *The Math Works, Inc.*, Natick, MA, 1999.

21. D. MAYNE AND H. MICHALSKA, *Robust receding horizon control of constrained nonlinear systems*, IEEE Trans. Autom. Control, 38 (1993), pp. 1623–1633.

22. W. MILLER, R. SUTTON, AND P. WERBOS, *Neural Networks for Control*, MIT Press, Cambridge (MA), 1990.

23. A. NAYFEH, *Introduction to Perturbation Techniques*, John Wiley, New York, 1981.

24. A. NAYFEH AND D. MOOK, *Nonlinear Oscillations*, John Wiley, New York, 1979.

25. K. NIKZAD, *A Study of Neural and Conventional Control Paradigms in Active Digital Structural Control*, PhD Dissertation, Dept. of Civ. Engrg., Univ. of Illinois at Urbana–Champaign, 1992.

26. K. NIKZAD, J. GHABOUSSI, AND S. PAUL, *Actuator dynamics and delay compensation using neurocontrollers*, ASCE J. Engrg. Mech. Div., 122(10) (1996), pp. 966–975.

27. T. PARISINI AND R. ZOPPOLI, *A receding-horizon regulator for nonlinear systems and a neural approximation*, Automatica, 31(10) (1995), pp. 1443–1451.

28. K. PASSINO AND Y. S., *Fuzzy control*, The Control Handbook, W.S. Levine, editor (1996), pp. 1001–1017.

29. W. PROCE AND R. BISHOP, *Probabilistic Theory of Ship Dynamics*, Chapman and Hall, London, 1974.

30. G. ROBERTS, *Ship roll damping using rudder and stabilizing fins*, Proc. IFAC Workshop on Contr. Appl. in Marine Syst., Genoa (1992), pp. 234–248.

31. D. RUMHELART, G. HINTON, AND R. WILLIAMS, *Learning Internal Representation by Error Propagation*, MIT Press, Cambridge, MA, 1986.

32. A. SAGE AND C. WHITE III, *Optimum Systems Control*, Prentice Hall, Inc., Englewood Cliffs NJ, 1977.

33. H. SCHUSTER, *Deterministic Chaos*, Springer Verlag, Weinheim, 1994.

34. M. SHINOZUKA AND C.-M. JAN, *Digital simulation of random processes and its applications*, J. Sound and Vibrations, 25(1) (1972), pp. 111–128.

35. M. SUGENO, *Industrial Applications of Fuzzy Control*, North Holland, The Netherlands, 1985.

36. R. SUTTON, G. ROBERTS, AND S. TAYLOR, *Tuning fuzzy ship autopilot using artificial neural networks*, Trans. Inst. Meas. and Cont., 19(2) (1997), pp. 94–106.

37. A. TIANO AND M. BLANKE, *Multivariable identification of ship steering and roll motion*, Trans. Inst. Meas. and Cont., in press (1998).

38. A. TIANO AND ET AL., *Rudder roll stabilization by neural network based control systems*, Proc. III Int. Conf. Manoeuvr. and Cont. of Marine Craft, Southampton (1994), pp. 33–44.

39. A. TIANO AND P. VENINI, *Fuzzy and neural controller of multivariable ship motion*, preprint, (1998).

40. A. TIANO AND A. ZIRILLI, *Fuzzy controller of multivariable ship motion*, preprint, (1998).

41. J. VAN AMERONGEN, P. VAN DER KLUGT, AND J. PIEFFERS, *Rudder roll stabilisation-controller design*, Proc. VIII Ship Cont. Syst. Symposium, The Hague (1987), pp. 120–142.

42. P. VENINI, *Neural Network Based Control of Simple Nonlinear Structures*, MSc Thesis, Dept. of Civ. Engrg., Univ. of Illinois at Urbana–Champaign, 1992.

43. ——, *Receding-horizon control of hysteretic oscillators (i): computations and numerical tests*, in preparation, (1998).

44. ——, *Receding-horizon control of hysteretic oscillators (ii): approximations by backpropagation neural networks*, in preparation, (1998).

Printed in the United States
By Bookmasters